Earth is an Engine

Written By

Michael B. Klocek

ISBN:

Hardcover: 978-1-962381-24-6

Paperback: 978-1-962381-23-9

Acknowledgment

Above all others, I would like to dedicate this book to GOD, the project manager of the creation of Earth.

Ron, "Hip! Hip! Hooray! Hip! Hip! Hooray!" and special thanks to Danea Young and Jackson.

Especially Demi, one of the smartest people on the planet, who told me to write this book.

Dedication

TO GOD

To the creator of our extraordinary home. The predominant construction manager of our Earth.

About the Author

Just a simple and observant person.

Contents

Page Blank Intentionally

Chapter 1: Agreements

The Hebrew Bible and the New Testament outline a belief in the existence of God. The book of Genesis presents that God developed the foundational framework of our Earth in just six days and proclaimed it to be good. The agreement that God's creations are good was established as one of the earliest agreements of our time. Additionally, the book of John in the New Testament states, "In the beginning was the WORD. The word was with God, and the WORD was God."

Consequently, one word and one agreement signify the commencement of an extraordinary world in which the beginning of time is regarded as an agreement. Based on Earth coming into existence, it seemed interesting to think about how time began, which was initiated with the ability to understand the fundamental mechanical apparatus known as Earth. When we examine the word "created," we will notice one of its synonyms could also be "constructed." One word, same meaning, different outlook. When one person stakes a claim to the construction of a tall skyscraper in New York by placing their name on top, we can be certain that they never picked up a hammer. Yet, people believe they created this monumental task.

Believing that our Earth was constructed also brings us to one word, one name: God.

When we think of agreements, they are what we learn, and they extend further than just written words on documents. When agreements between societies began, they consisted of many words and propositions that made communication easier and faster. Then, the agreements were narrowed down to one word. For instance, in order for communication to be understood at a quicker pace, societies agreed to convey propositions using just one word. A prime example is the description of animals, where each creature is represented by a single word. No matter what language, societies agreed on describing animals with one word. Noticing what is being described, understanding becomes easy and fast.

Throughout the world, we employ one-word agreements encompassing terms like 'dog,' 'cat,' 'horse,' 'hog,' 'donkey,' and countless others. This principle extends to buildings and objects as well; they are encapsulated into single words. Since societies agree on just using one word, we can easily see how they are considered agreements. These agreements are used for easy understanding and to make communication faster.

Subsequently, words are combined to form sentences which enable us to communicate. As we progress through life, we accept learned words, creating agreements.

Examples encompass everyday items like clothing, our places of residence, and the food we consume. Behind these agreements are definitions. We accept and agree that doors are used to enter and exit, cars are used for transportation, and school buses take kids to school. Then, we navigate through life, employing words that denote the subjects of our discussions, such as 'bus,' 'car,' and 'house.'

Laws of the land, laws of society, laws of gravity, and laws of circular motion are agreements that are not learned, just accepted. When we set our foot down on the ground, we learned and accepted gravity at a younger age. However, we have never learned how gravity works. Through experience, we formed an agreement that gravity exists. Therefore, certain things eventually became acceptable through experience. Societal cultures and agreements are also learned, such as holding a door open for someone. Proper communication and understanding are based on agreements. It can either be the case that communication flows smoothly or becomes immensely complex. Disagreements take place when words are spoken in a language that one or both parties don't understand.

Having a clear and to-the-point agreement is a vital component of any discussion or dialogue. A lack of understanding or an incorrect interpretation can cause disagreements and can result in outrage. Misunderstandings emerge when people don't mutually agree on the meaning of a word, resulting in confusion, arguments, and even chaos.

A person's language, experiences, culture, and background sometimes result in words and agreements that have different meanings. For the purpose of maintaining social synergy and proper communication, it is vital to come to agreements. Also, we all must mutually agree on the meanings of words as well as their definitions. Throughout history, there have been examples of inaccurate agreements and profound consequences that can result from such unshakeable beliefs.

A Polish astronomer named Nicolaus Copernicus tried to prove to the world that the celestial bodies of our solar system revolved around our Sun by writing and publishing 'On the Revolutions of Heavenly Spheres.'

Still, the world found it hard to unite together and form an agreement. With no agreement, the world of education about this subject was set back another century. One more notable example is the experience of Galileo Galilei, an Italian astronomer whose groundbreaking discoveries challenged the prevailing agreement that the Earth was situated at the center of the universe. Galileo's observations and experiments caused him to support the heliocentric model, which positioned the Earth, as well as other celestial bodies, revolve around our Sun. This progressive proposition opposed the vastly acknowledged geocentric model, which maintained that the Earth was fixed and any remaining astronomical bodies revolved around our Earth. Galileo's heliocentric theory directly opposed the religious and philosophical beliefs of the time, even though it was backed up by compelling evidence and scientific truth.

Supported by religious principles and influential people of authority, the agreement that our Earth was the center of the universe was recognized by society for a very long time until Galileo proved it to be incorrect. This assumption that the Earth was the center of the universe was a very solid agreement that was based on false information and misinformed education. When Galileo aimed his telescope into the night sky in the year 1610, he observed for the first time in human history that celestial bodies orbited Jupiter.

Also, if Aristotle was right about the geocentric model, which implied that everything in outer space orbited the Earth, then these moons could not have existed. Galileo also observed the phases of Venus, which showed that the planet orbits our Sun. At around the same time, German mathematician Johannes Kepler was in the process of publishing a series of laws that illustrated the orbits of the planets around our Sun. Finally, in 1687, Isaac Newton was able to prove that Aristotle's geocentric view of the universe was invalid by illustrating that planets moved around our Sun and that 'gravity' was the force that kept them in check. We will delve into the study of gravity later in the book. Proving that gravity on Earth is mostly circular, providing us with a new agreement.

Galileo encountered a lot of resistance as his ideas spread and challenged the consensus. The framework that was utilized at that time declined to even consider the validity of what other astronomers saw. Instead, they held on to the geocentric view with absolute disbelief of change. Galileo's findings were met with severe opposition, eventually prompting his judgment by a Roman Catholic Church investigation. For supporting his heliocentric views, Galileo was accused, then judged, for holding a contradictory viewpoint. Then, in 1633, he was placed under house arrest for more than three decades, compelled to hold back his ideas, and was separated from his family and the scientific community.

In spite of presenting solid evidence to support the truth, this incident serves as a clear reminder of how deeply ingrained and thoroughly structured agreements can become.

The case of Galileo serves as an example of the enormous power that is present between languages, countries, and societies. They have the power to direct social and political structures, shape collective beliefs, and even control the direction of scientific advancement. It is likely for agreements to restrain innovation, halt the pursuit of knowledge, and enforce false narratives when they become rigid and opposed to change. Fast forward ahead, and we can examine Darwin's theory of evolution, noticing that it has set the advancement of the academic and scientific communities back hundreds of years. Leaving many aspects of life unobserved.

By acknowledging that Earth itself is a mechanical engine – a dynamic and interconnected system that functions on various levels, we can reset agreements. We can do this by changing what has been agreed to in the past and starting a new beginning. Without a doubt, a profound and new understanding of how our planet functions, spins, retains its orbit, or even how it exists in our solar system has to start **somewhere**.

Let's start with the agreement that the Earth is spinning counterclockwise. Then, add to the agreement that this overall counterclockwise spin is being caused by something mechanical. Also, we can agree that the Earth tilts back and forth on its axis, as the Earth revolves around our Sun.

In viewing the Earth as a mechanical engine, we realize that it is not just a passive thing, but instead, it has many complicated systems that work together with an opposed balance. These all-encompassing systems enable us to understand how separate engines, combined as one, interconnect and produce many effects. These intertwined effects, which are being produced, show how Earth exists.

There are numerous interconnected parts and levels that form Earth's Engine, and each one is important to how the planet functions. It is the Earth's internal dynamics that drive geological phenomena like ocean currents, squalls, typhoons, tornadoes, Earthquakes, volcanoes, and hurricanes.

The Earth's Engine maintains a delicate mechanical equilibrium. Understanding these interconnections through agreements provides us with a significant amount of knowledge about the functioning of our planet, as well as how human activities affect the mechanical engine's stability. In order to better understand how to maintain a mechanical equilibrium, let's consider how Earth's temperature depends on the balance between energy entering the electromagnetic field plus the energy being released from inside the planet. For instance, when incoming energy from our Sun is absorbed, Earth tends to get warmer.

When energy from the mechanical engines of Earth is added to the bombardment of energy from our Sun, these combined energies must be cooled. Our Earth must be protected from overheating. When the balance between hot and cold wavers from one extreme to another, the planet becomes unlivable. This extreme shift has significant repercussions for quite a few research fields.

Trying to solve problems by studying effects rather than focusing on the cause of these catastrophes will only solve half of our problem. Catastrophes should energize a joint research effort by studying the cause and focusing on how to create cooperation between two completely different sides. One side is violent and destructive, causing disasters and a 'State of Emergencies,' and the other side pushes back, as it forms a more tolerant place to live.

To move forward and find a solution to the challenges we face as a society, we must willingly accept the notion that Earth is a mechanical engine. Once we come to the realization that our Earth is an active and dynamic mechanical system, we will automatically acquire the skills and knowledge to solve the problems we encounter. At its very core, this understanding holds the agreement that Earth is the cause, not just an effect. For this reason, we should acknowledge that the phenomena we observe and experience, such as the climate, stem not from outside powers such as 'Mother Nature' but rather originate from within our Earth. This agreement opens up the possibility for a new era of investigation that was previously acknowledged but not fully accepted.

Certain agreements highlighted in the book of Genesis and the New Testament emphasize the crucial role they play in communication and societal function. Words are vital for communication, and agreements are necessary for societies to function at a reasonable pace. We must agree on definitions and recognize the significance of learning individual words. Additionally, we should be able to understand words in order to communicate effectively. However, we also should be willing to change our agreements when new information comes into our perspective. The new agreement presented here acknowledges that we are a part of this interconnected mechanical system. When we agree that our Earth is a mechanical engine with nuts and bolts, it is our obligation as human beings to keep Earth's mechanical engines in a state of equilibrium and balance. We must have an understanding of our impact on the environment since our actions have consequences. Fully accepting the agreement that Earth is an engine propels us to take our responsibilities seriously and go about the future as stewards of our planet.

A friend of mine texted me after reading this chapter on agreements. He said that after he read it, he looked around his office and started noticing things he never realized how they came into existence. He noticed how his clock revealed time, the carpet, his shoes, a pen, the little cup on his desk to put pens and pencils in, and his computer and paper all started with an agreement or a need. He became aware of his surroundings.

Countries, religions, and all creeds must accept the agreement that Earth is a mechanical engine. Upon completion of this book, just like my friend, you will become aware as well. Afterward, all of us will be able to facilitate a deeper understanding of how we can unite. For instance, we can notice the presence of an underlying cause. By studying what is at cause, we can gain a better understanding of the planet and its inner mechanical workings.

Soon after embracing the agreement of our Earth as a mechanical engine, we will observe and change difficult truths. These truths will work as a reminder to us that, as living beings on this planet, we are liable for maintaining the health of the mechanical engines of Earth for future generations. By wholeheartedly creating a new agreement, we can cooperate and grasp the complexities of this reality and guarantee its continued compulsions. The agreement that our Earth is a series of mechanical engines has to initiate from somewhere. This agreement opens the door for investigation, understanding, and the improvement of a common vision for our planet's future.

This new beginning will lead us to new discoveries. In the next chapter, we will determine how agreements become solid and difficult to change as we elaborate on the reality of 'Increments and Gradients.'

Chapter 2: Increments and Gradients

Increments and gradients play a significant role in shaping our understanding of reality. From the formation of agreements to the development of education, technology, and societal norms, increments and gradients influence various aspects of our lives. Agreements serve as the foundation upon which our beliefs and perceptions are built. We agreed at one point in our human society that we should reside indoors to protect ourselves from the weather. Then we built a shelter or a house. Then we agreed that the house should have a door, then a handle to open and close the door, then we agreed to call it a door knob and installed a barrier or lock. Increments consist of a certain grade or gradients. It's how we learn, teach, grow, build, and most of all, they are part of how we live. When we look around, we can observe that every object in our world originated from an agreement. Whether it's automobiles or furniture, each entity emerged from an initial agreement that formed out of a need.

Then, gradually, the agreement stayed the same over time. Historical photographs from the past provide glimpses into the transformation of objects and structures, thus capturing the enhancement of a solution by fulfilling a need or creating a better way of life, especially when we have a need to solve problems.

The beginning of a need for faster communication progressively allowed us to have cell phones thousands of years later. When an inventor finds a solution to a need or invents something completely new, we teach young generations these inventions as simple agreements or words, such as a toaster, shower, and microwave. Societies all around the world commonly use one or two words to describe these new agreements. Such as cell phones, iPads, GPS, and abbreviations, such as SUVs, ATVs, WWW, .com, and especially phrases, such as "text me." This phrase used to be, "Write me," then it became, "Call me." Tracing back through time when communication primarily consisted of letters and books chiseled onto stones.

Ask someone today, “Let’s saddle up the horses for a visit to grandma’s house.” and the confusion would set in. Foreseeing the future of science, the agreement of having a grandma one day could be just as obsolete as saddling up a horse for a visit. Observing the construction of our Earth can be seen as an agreement made by a divine entity, setting in motion a gradient that defines the purpose of our world.

Craftsmanship also reflects the influence of increments and gradients, as the materials and skills available dictate certain agreements. For instance, in regions with a limited supply of wood, or wood was used for heating and cooking, tables were made of stone. Now, we have all types of tables made from different materials, with different names or agreements, such as coffee tables, end tables, and dining room tables made from plastic, wood, and other materials.

Increments and gradients also shape our personal experiences. Think back to the first time you learned to tie your shoes or make a knot. Initially, at a young age, it was difficult, but then the task became rather simple through learning and practice. These examples emphasize the importance of gradual progress through learning and practice. Education is another area heavily influenced by increments and gradients. Students progress through various grade levels, gradually building knowledge and skills. Missed increments and skipped grades can lead to difficulties in understanding, potentially causing students to disengage or drop out.

Effective educational approaches focus on identifying misunderstandings, solidifying agreements, and adjusting the gradient to ensure learning and growth. Increments and gradients also leave a significant imprint on technology and societal enhancements through progression.

Technological advancements, for instance, are driven

by agreements and learned through increments and gradients, which provide solutions to our problems. From massive computers that once filled entire rooms to the compact and powerful computers we hold in our hands, technology has undergone significant transformations due to learned increments and gradients. Similarly, societal norms led to things being forgotten or obsolete. For example, black and white televisions have gradually been replaced by vibrant, high-definition.

Technological advancements progressed through competition and profit. Increments and gradients are intertwined with the passage of time, and they establish all things. Life cycles of living beings, or even man-made, follow gradients of growth, stability, and decay. They involve birth, aging, and eventual passing, while the things we create require maintenance and repairs to delay their deterioration.

Ultimately, everything follows a downward gradient, paving the way for new beginnings and enhancements. An example is the progression of music, from vinyl records to eight tracks and cassettes to digital files and now streaming services through Wi-Fi. This exemplifies the ever-changing agreements, increments, and gradients within cultures.

As we delve into the contents of this book, it is important to recognize that our understanding begins on a lower gradient scale within the initial chapters. Then, as we progress, our comprehension deepens, agreements become concrete, and we climb to higher levels of understanding; through small increments, our agreement that Earth is a mechanical engine will become apparent. Increments and gradients are integral to our lives. They influence our learning by showing us how agreements were developed, including education, technology, and societal advancements. They serve as the mechanisms through which we understand our world.

By appreciating the role of increments and gradients, we gain a deeper understanding of the gradual progression of one state to another. Every object we encounter, every agreement we form, and every notion we grasp is built upon growth and development. These agreements shape our realities, influence our perceptions, and guide our interactions with our world. Understanding the importance of gradual progress and recognizing the potential consequences of disrupting agreements on a too steep of a learning curve are important to moving on to the next level. Increments and gradients are agents of change by reflecting the passage of time and the existence of all things. Life itself follows a gradient, with birth, growth, and eventual decay.

Similarly, the things we create, from buildings to technological advancements, undergo cycles of maintenance, repair, and eventual obsolescence, such as wheels made out of stone, then wood, and, eventually, bright and shining chrome.

Increments and gradients illuminate the progression of our reality, reminding us that everything is subject to change. The journey through education highlights the significance of increments and gradients as well. Students progress through various grade levels, gradually acquiring knowledge and skills. Missed increments and gradients can slow understanding and impede progress. It is essential to identify and address these gaps, allowing for a comprehensive and successful learning experience. Technological advancements exemplify the impact of increments and gradients on society.

As agreements evolve, new technologies emerge, reshaping our lives and altering the way we interact with our world. Society itself is governed by agreements that are enhanced over time, reflecting changing values, beliefs, and societal norms. To navigate the ever-changing foundation of agreements through gradients, we must foster an awareness of their influences. By recognizing their existence and appreciating their role in our lives, we can develop a deeper understanding.

In summary, increments and gradients are fundamental to our understanding of reality. They shape our beliefs, influence our interactions, and guide the development of our society.

By embracing increments and gradients and the agreements they form, we can navigate the complexities of our ever-changing world with wisdom and insight. As we examine and investigate the spinning of our Earth, we agree it revolves around our Sun. One of the agreements for the existence and function of our Earth is 'Mother Nature.' This agreement of the unknown has been left alone, with no further examination or investigation, until now. On an increment and gradients learning curve, we will reach a level of understanding that Earth is a mechanical Engine. Now, moving on to the next increment, we have Dichotomies.

Chapter 3: Dichotomies

Dichotomies represent the division or contrast between two things that are opposed or opposite. Dichotomies are present in every aspect of life, from the climate of our Earth to human societies. They contribute to the progression and advancements of technology. Examples of dichotomies include wet and dry, hot and cold, hard and soft, positive and negative, fast and slow, light and dark, good and evil, God and the devil, winners and losers, clockwise and counterclockwise. In fact, most things that exist are dichotomies. They exist in opposition to each other, yet together, they create balance and equilibrium. The way this balance works is how most things are controlled.

An engine would spin out of control and run forever without the dichotomy of a functioning transmission. Engines need something to slow them down and maintain a safe balance, whether through brakes or a properly operating transmission. Engines, just like most operating entities, need to have a balance between hot and cold. Living things function within specific temperature ranges; for humans, this range typically falls between ninety-six to ninety-nine degrees. The dichotomy of wet and dry also needs balance. Place an engine in water for too long, and it won't operate properly. If an engine dries out, friction will eventually cause it to overheat. Engines need something to keep them warm and cool. Just like living entities use water and liquids to keep them cool, engines use oil, liquids, and wind. Therefore, dichotomies extend farther than just being opposed to one another; they go beyond mere opposition between two extremes. They require a state of balance and equilibrium in order to have stability.

For example, a balanced dichotomy of hot and cold is essential for our survival as humans. When we get too hot, we die. When we get too cold, we die. It's the same for engines. This is not a philosophy but a reality of life.

With little or no oil in the engine of your vehicle, during winter, your engine will freeze and won't start until you warm it up. Furthermore, dichotomies do not have fixed or absolute limits as long as the dichotomy creates balance with equilibrium. For a dichotomy to exist, one side has to be greater than the other. Otherwise, there will be stagnation. Usually, the positive side, or input, is the side that has to be controlled. Thinking about how the balance of hot and cold can be applied to the proper functioning of engines, we notice that engines operate at extreme temperatures. How the balance between hot and cold is maintained in machinery is a simple answer: oil, water, wind, and exhaust.

Maintaining the hot-cold balance is essential for engines to operate at their optimum. The engine can only operate effectively in a variety of conditions because of oil, water, wind, and exhaust.

Moreover, dichotomies establish balance and equilibrium as they operate in opposition to each other. Remember, one side must hold greater influence; otherwise, stagnation occurs. Usually, the positive side is more powerful than the negative.

In conclusion, dichotomies are an essential part of the world. They represent the division or contrast between two things that are opposed or entirely different. At the same time, this opposition creates balance and equilibrium. While they can be useful in helping us understand our Earth, it's important to recognize that they have no limitations and exist in a delicate balance by keeping each other in equilibrium. By understanding this delicate balance, we can appreciate the advanced engineering needed to create this balance, also agreeing that this balance couldn't possibly be formed by an unknown source such as 'Mother Nature.' Therefore, it becomes obvious that dichotomies are designed.

Now, we have examined agreements, followed by increments and gradients. Designed dichotomies begin the journey of realizing the fact that Earth is a mechanical engine.

Let's take this journey one step further and move on to the next adventure by examining how things work with the next gradient: “Functions.”

Chapter 4: Functions

In this chapter, we will explore the notion of functions and how they apply to understanding our world. Our agreed-upon premise is that our Earth is an engine. In order to comprehend how it operates and how we can improve it, we must examine the functions of its various components. Let's begin with a fundamental question: *What is the function of leaves on plants and trees?*

Science tells us that leaves absorb Sunlight and produce chlorophyll, which, in turn, provides nutrients for maintaining the life of plants and trees.

As they grow, they absorb carbon and produce oxygen. However, leaves serve a more crucial role. In addition to producing oxygen and absorbing carbon, leaves, and groundcover also provide shade. Since we are focusing on the fact that Earth is a mechanical engine, we know that engines need to remain cool to operate effectively. When something begins to spin, such as an engine, friction causes it to heat up. The compartments of engines are designed so that when they are moving, wind flows over the engine to keep them cool.

Therefore, when wind passes through trees, it cools the surface of our Earth. Can you notice the diverse functions of trees and ground cover beyond just producing oxygen and absorbing carbon? The absorption of carbon by trees and groundcover is also part of the exhaust system of Earth. Studies show that when combined, all trees create a higher and cooler production of wind, and the wind passing through the trees picks up pace in a continuous motion, cooling the surface of the Earth. Providing shade, trees, and ground cover are some of the first defenses designed to keep the engines cool from the bombardment of heat from our Sun.

To cement our agreements, let's look at other examples of functions. For instance, oil is broken down into different types of substances and lubricants, which prevents moving parts from overheating. It is also used to create various fuels, lotions, and personal care products, such as Sun tan lotion and hand creams, which shield us from dryness and heat.

Water also has numerous functions. The primary function of water is to keep things cool. Nuclear engines, for example, would melt down and explode without water. We drink water to stay cool or to hydrate. Water also acts as a solvent; try washing your clothes or your car without water.

However, an overlooked or unseen function of water is that its combined weight acts as a stabilizer, preventing the

Earth from wobbling. Water and liquids flow into a path of least resistance, forced by gravity, causing balance. When combined, water is the heaviest substance on the planet and acts as a counterweight, keeping our Earth from wobbling out of control. Also, the combined weight of water produces a pressurized cage, stabilizing the apparatus and engines. We can notice the different functions of certain entities are right before our eyes, often unnoticed and taken for granted.

We must focus on analyzing the functions of our world and stop relying on vague notions like 'Mother Nature.' Instead, we must study the functions through simple observations of our surroundings in order to find ways for improvements. We must understand that everything has a function, and most exist within a dichotomy. Climate change is a complex topic that we will delve into, along with its solutions, later in the book. For now, let's concentrate on exploring functions and discovering what makes things work and how they can be improved.

By focusing on a low gradient at a medium speed, we can examine the functions of major entities like our Moon, our Sun, and gravity. Once we comprehend their actual functions, we can identify their common traits and reach an obvious agreement.

Keeping the engines cool, stabilized, and *lubricated is our overall goal.*

By staying on an even gradient, we can move forward and discover the cause and effect of almost everything.

Gradients should progress smoothly, allowing us to navigate through complex functions without going too fast. An even gradient helps us avoid abrupt shifts and enables us to appreciate the intricacies of observations. Just as a transmission with gears, an even progression through gradients permits us to transition smoothly between various distinct functions. This, in turn, provides us with a clearer picture of its behavior, whether it's trees, water, moons, or oil. Such an approach leads us to a more profound comprehension of the underlying mechanical engines.

For instance, dragsters are designed with only two gears, forward and high, and are built to unleash the power of their engines just once per race. Their setup includes a centrifugal clutch, emphasizing a focus on input over output, specifically the use of jet fuel. Without any opposition or the need to slow down, dragsters reach high speeds swiftly, using a parachute to slow down, and once the race is completed, the engine of the dragster doesn't function properly anymore. The engine needs to be rebuilt because it was designed to go fast only once. When engine designers add an opposition, such as a transmission with gears, the engine can be used continuously.

For instance, when skiers go down the side of a mountain, in order to stay in control, they go from side to side, which slows down their speed, providing for a balanced gradient. Therefore, the effects of certain functions are obvious yet go unnoticed.

In conclusion, understanding the functions of the obvious is critical to the comprehension of our world. From plants and trees to water and oil, everything has a purpose. Our agreements normally exist within a dichotomy. By focusing on functions, we can improve our world. With these improvements, we can keep the mechanical engines of Earth cool, stabilized, and lubricated. Next, we will examine the simple principles of movement - 'Cause and effect.'

Chapter 5: Cause and Effect

Cause and effect are fundamental notions that govern the workings of the universe. It's a simple proposition - whenever there is an action, there is a reaction. Things become a little more complicated when we introduce the structure of a dichotomy. Along with every action, there is a counter-action, and this creates movement and balance, and it's a reality that applies to everything in our world.

For example, if we push something, it moves. When it pushes back, we have a dichotomy - a counter-action or opposition. Without this counter-action, it would move continuously, eventually spinning out of control. In this dichotomy, we have positive and negative forces, with one pushing forward and the other pushing back. This creates balance and equilibrium, allowing us to move forward while keeping everything from spinning out of control.

Dichotomies are everywhere, from the push and pull between criminals and the police to the battle between good and evil. Even the natural world operates on this principle, with hot and cold, life and death, all working together in a delicate balance. In order to keep equilibrium and still move forward, one side of the dichotomy has to be slightly greater than the other. Positive, good, God, life, peace, and love are all examples of the sides that usually win out. These realities create a steep gradient, which allows us to move forward and make progress.

We live in an ever-changing world where cause and effect, action and reaction, and action and counter-action are constantly at play. Most actions and reactions are easily noticeable; we can see and feel them. When we press down on a gas pedal, the engine revs up. When we turn on a faucet, water comes out, and when we call someone a name out of hate, we are not going to get love in return.

A fun example of cause and effect is watching a video of an inchworm.

There are also larger realities at play, such as the spinning of our Earth. We can feel it, and we can see it in the skies. Yet, we often use vague concepts like 'Mother Nature' as an explanation. The truth is that **something** is causing the Earth to spin, and it is all governed by cause and effect.

As we continue on our journey, we will explore the mysterious realm of Earth and dive deeper into the principles of cause and effect. We will see how these principles apply to everything in our world and how we can use them to make progress and create positive change.

Remember, without cause, there is no effect, and

without action, there is no reaction. Without counter-action, there is no balance, stability, or equilibrium. Let's recapture the lessons we have learned so far. We have agreements that are developed and learned over time. When small steps or increments are combined, they create grades or gradients.

Dichotomies are formed, which create balance and equilibrium. Functions are studied through cause and effect. Next, let's put these lessons to good use. Any gradient can take some time to become solid or real. There are many questions that need to be answered for us to figure out what our Earth is and how it functions.

Chapter 6: Our Earth

Let's begin a new gradient by looking into the book of Genesis, where God created our Earth in six days. Now, it becomes evident and obvious that days, as a twenty-four-hour time frame, didn't exist until the Earth was completed and ignited. Therefore, comprehending the duration of a day before our Earth began running at full speed, time was measured differently. In fact, sixty seconds in a minute, sixty minutes in an hour, twenty- four hours in a day, and three hundred and sixty- five days in a year only exist on planet Earth.

Is it hard to believe that over an unknown period of time, the Earth was engineered by design, then constructed, and not mysteriously formed? Honestly, which makes more sense? Over time, our Earth was formed by a make-believe creature called 'Mother Nature,' or over time, our Earth was engineered and constructed as a mechanical engine.

Let's also look at how great societies were not only built they also slowly diminished.

The word, 'independence' floated through the Colonies of America for some time. Then, the colonists took the next gradient and caused the word 'independence' to become real and solid. The United States of America is the result or effect of the agreement called 'independence.'

On the reverse, the longest and greatest civilization on Earth, the Roman Empire, slowly eroded away, and new societies were built. Through observations, new gradients are being started every day while old agreements are being replaced. The same is the case with our Earth. We orbit counterclockwise around our Sun every three hundred and sixty-five days. This complete cycle is called a year. As we spin counterclockwise, every twenty-four hours completes a cycle, which we call a day, twelve hours of light and twelve hours of darkness. These timetables of minutes, hours, days, and years are effects. Weather, Earthquakes, and lightning storms are effects, too.

Most everything we see, feel, and experience on the surface of our Earth are effects caused by **something** located within our Earth. Earth is the only solid specimen within our line of sight that we can dissect and study. Looking for a cause, we can track the events of effects or reactions by simple observations. Yet, the agreement is that Earth is a planet derived from an extremely vague concept called 'Mother Nature.' This make-believe entity has a tendency to lead scientists and students astray. It is very doubtful that 'Mother

Nature' created our world. Yet, this is the agreement; ask any scientist. This debate goes on with unexplained concepts of what is in the core of Earth and what gravity is. The answer from scientists varies; therefore, we don't know exactly. But, of course, *we* know. It's a mechanical engine. **Something** is at cause. The complicated and simple functions of our Earth create multiple effects, such as huge land masses, the movement of ocean currents, flows of volcanoes, and the unpredictable effects of hurricanes and tornadoes.

Questions, such as how and why our Earth tilts back and forth on an axis, are explained with vague propositions. We have winter in the Northern hemisphere, and at the same time, summer is taking place in the Southern hemisphere. These effects are being caused by **something**. Let's reconsider the belief and change the agreement that the existence of our Earth is a mythical entity called 'Mother Nature', and instead create a new definition, a new agreement. Earth is a mechanical engine.

We need to examine this new agreement slowly. A good example is when Jesus got into trouble and eventually ended up on the cross by going too fast on a learning curve. Proof of this is at the age of twelve, he had more knowledge about the Bible than all others.

His fast rise to power in his thirties forced the fear of the unknown across societies. This steep gradient caused his inevitable demise.

By going too fast on an unacceptable learning curve, the same demise will happen.

Earth is a mechanical Engine. But how do we know? To explore this, let's start and compare our Earth with an engine, such as the one in an automobile. The Earth's engine is designed completely differently than automobiles, trains, planes, or ships on the surface of Earth. In order for Earth's engines to spin, they also need compression. The compression of Earth's engine is produced by electromagnetism, centripetal force, and the pressure of water.

These forces are also being held together by an electromagnetic field and an electromagnetic cage.

These entities will be discussed in the next chapter on a steeper gradient. These forces within our Earth are extremely hot. The cause of centripetal force and electromagnetic force creates torque. Torque produces friction; friction produces heat. In order for an engine to continue working, the engine needs to stay cool. What keeps engines cool? Oil, water, wind, and exhaust.

What exists on our Earth that keeps its engines cool? Oil, water, wind, and exhaust. As you read this chapter, you will embrace a new gradient. As the gradient intensifies, the agreement becomes stronger, more tangible, and real. Gradients can sometimes be steep. Therefore, let's review it once more.

Our Earth is spinning counterclockwise. This we know, with no doubt. How our Earth is spinning will be the subject. The counterclockwise spinning of the Earth and gravity are formed agreements. Meaning that the scientific communities agree that both are taking place. These are true statements. Both are effects of **something**. The scientific community also agrees that our Earth tilts back and forth on an axis; this is also an effect. Give or take a few days, our Earth tilts back and forth, completing this cycle around the same time every year. These are called the winter and summer solstices. Scientists agree, due to video observations of our Earth from space at the North Pole, that our Earth is spinning counterclockwise.

When viewing video straight down at the South Pole, scientists agree that our Earth appears to be spinning clockwise. Also, this brings us to the middle of Earth, called the Equator.

The Real Equator

The equator represents one of science's worst misconceptions. This man-made, false line that divides our Earth from north and south is one of the causes of studies and research being led astray. The equator is a man-made line that deceives the world, especially young students, about the correct division between north and south.

One thing I would like for the readers to understand is that the center of our Earth is not stable. The division between North and South is not divided on the outside; this division is being created from the inside, at the center of our Earth.

The center of the apparatus wobbles and fluctuates as it floats and spins. This constitutes one of the main reasons for the paths of hurricanes and tornados. The forces of the main engine of the apparatus generate effects along the real equator. The center of the apparatus and the core are not precisely stable; they fluctuate within a certain range. This will be elaborated further in the next upcoming chapter titled "The Engines." However, for now, comprehend that the center engines spin, causing our true equator to wobble in a gyroscope-like effect. Moreover, when the northern sets of grinders slow down, and the southern sets of grinders speed up in opposite directions, these opposing forces create a wobble that impacts the entire apparatus. This wobble, or fluctuations, create results of studies that are extremely difficult to comprehend where the real equator operates.

The real equator, when tracking on the surface, can be seen along the areas of our Earth that are deserts and jungles. Let's start with the major deserts.

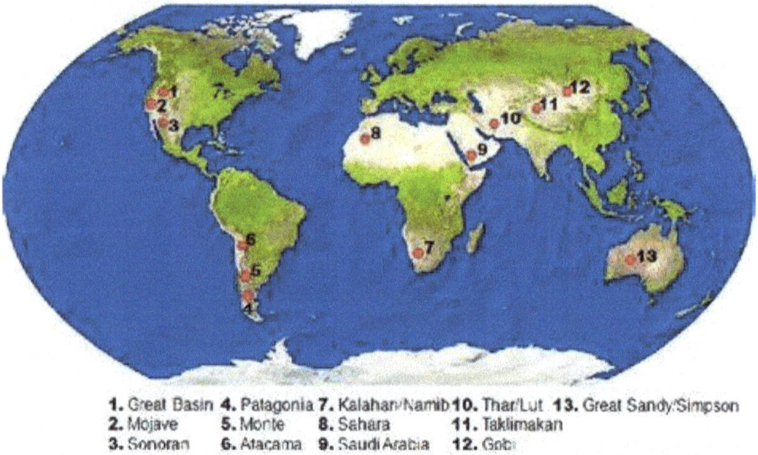

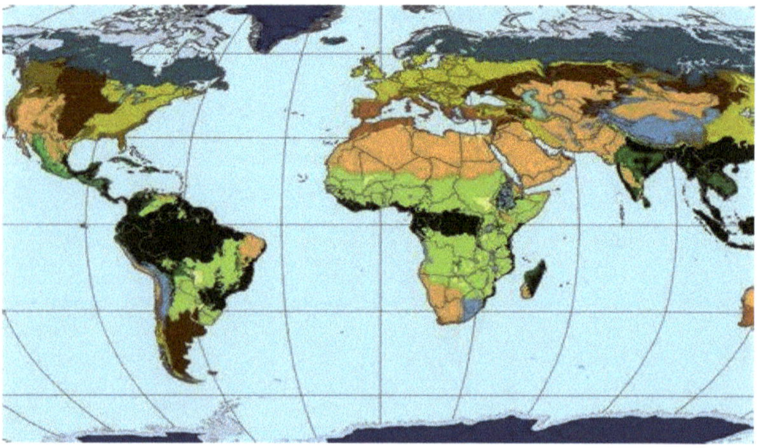

By drawing a real line through the locations of these deserts, we will be able to examine an outline of the real equator. Start the line in the Sahara Desert of Africa since it is the largest desert outside the Arctic poles.

We will travel the line counterclockwise since we are spinning in that direction. Continue the line from the Sahara through Saudi Arabia, then the Thar/Lut, follow into the Taklimakan, and then the last desert, the Gobi, which completes the circumference. This is close to the real equator we can witness on the surface of our Earth, remembering the real equator is wobbling inside Earth at the core.

To break things down into simplicity, when we want to follow the fluctuation of the center apparatus from space, just follow a map of the deserts and jungles.

Look at this map of deserts and just above the vast jungles, draw a line through any of them, and you will find the fluctuation of the center engine or the real equator. While you follow this fluctuation, you will find the trajectory of the center core engines. Other deserts of the Earth, such as the Mojave, follow the trajectory of the grinders. The grinders are part of the apparatus, which will also be discussed in the next chapter. This completes the unraveling of the man-made, imaginary line created by scientists called the equator.

A fun fact at this time is to notice the difference between the habitats of jungles and deserts. Notice something familiar? They form a dichotomy, where one is hot and dry, the other being cool and wet.

Once again, when observing our Earth's video directly from space, scientists agree that our Earth at the North Pole is spinning counterclockwise. Scientists also agree with the video that depicts our Earth viewed from directly above the South Pole, showing an apparent clockwise rotation. These two effects also create a dichotomy. Studying graphs of ocean currents also shows proof that a dichotomy of our Earth spinning counterclockwise in the North and clockwise in the South is consistent. How about the agreement of scientists that our Earth was formed and not created? Our Earth was not formed; it was constructed. Proof of this statement is based on these observations. Look around you. Does it seem as though our Earth just magically formed over time? The statement 'over billions of years' is used quite often in the scientific community to explain uncertainty. When scientists are uncertain, they use phrases like 'Mother Nature' and 'over time,' never explaining who Mother Nature is or the exact time.

In the chapters ahead, we will form an agreement to get us close to certainty. Nothing is absolutely certain. Nonetheless, by the end of this book, we are going to get closer to certainty than ever before. As I wrote earlier, one hundred percent perfect and complete certainty doesn't exist.

This includes myself.

So, we have to do this together. Do we have an agreement that Earth is a mechanical Engine? Not yet. The next chapter begins a journey into the center of Earth.

Can you handle it?

Are you ready for a steep gradient?

The hardest part of this book is next. Prepare yourself by taking notes. Let's journey into the unknown by learning about; 'THE ENGINES.'

Chapter 7: The Engines

The star of David, plus trillions of other brilliant stars that light up the night sky, are harmless to our Earth. As we look up into the sky at night and ponder the stars from our distance, we see beautiful glowing lights.

Through numerous telescopes, close-up views show beautiful pictures of different spheres with extravagant names, such as quasars, red giants, and pulsars. At a closer distance, they pose as much of a threat to our survival and the existence of Earth as our own Sun.

Developing our agreement that Earth is an Engine, here are the ideas and theories scientists put forth. Planets are formed by smashing into each other, causing different gravitational pulls. Then, after billions of years, they formed a center of collected metals, and the metals reformed into fire. Then, an unidentified entity called gravity spun the planets into different shapes, forming distinct celestial spheres.

Asking a simple question at this point leads extremely intelligent people to fall into confusion.

"How?" Another question is, "Can you explain gravity?"

The agreement of a false entity called 'Mother Nature' is the reason further investigations and studies on these subjects have halted. Why do scientists relinquish their aptitude to something that is unidentified? Simply put, they moved on to easier explanations without really answering the questions. Another good question to ask is, "Why are planets and moons different, with different functions?" Our sun heats things up, while other planets and moons contain gas and water, such as Jupiter, Neptune, and Triton.

Also, some moons spin out of control, while others are quite peaceful. Some planets look and seem peaceful, yet the gravity of Jupiter is so immense that the human race cannot sustain life.

As we watch delegated science programs or read the scientific theories on how planets form, they create extreme difficulty to comprehend. Explanations of these theories become monumental as scientists try to explain a believable basis for their philosophy. When there is truth and reality behind an agreement, it becomes very easy to explain through observations.

In this chapter, we will examine some well-known observations. These observations are the results of something at cause, which is creating effects. These effects will help prove the existence of the apparatus, which contains the engines of Earth. Here is a list of some very recognizable effects:

Clouds, lightning, volcanoes, tornadoes, underwater lava shoots, ocean currents, jet streams, hurricanes, tornadoes, mountains, and humidity/steam/fog. By recognizing the effects above, it becomes easy to realize that **something** is at cause. From here until the end of the book, starting with the largest engine, which creates the center of the apparatus, we will work our way out, step by step, increment by increment, grade by grade. Again, we will work our way from inside of Earth, from the center of an engineered and constructed apparatus. Then, continue all the way out to examining the stabilization of our Sun. There are five entities that an engine needs to operate:

1) Oil

2) Water

3) An electrical system

4) Wind

5) Exhaust

Keeping these functions in mind is crucial to understanding the Engines of Earth. Let's begin with a general description of the entire structure called the apparatus.

The Apparatus

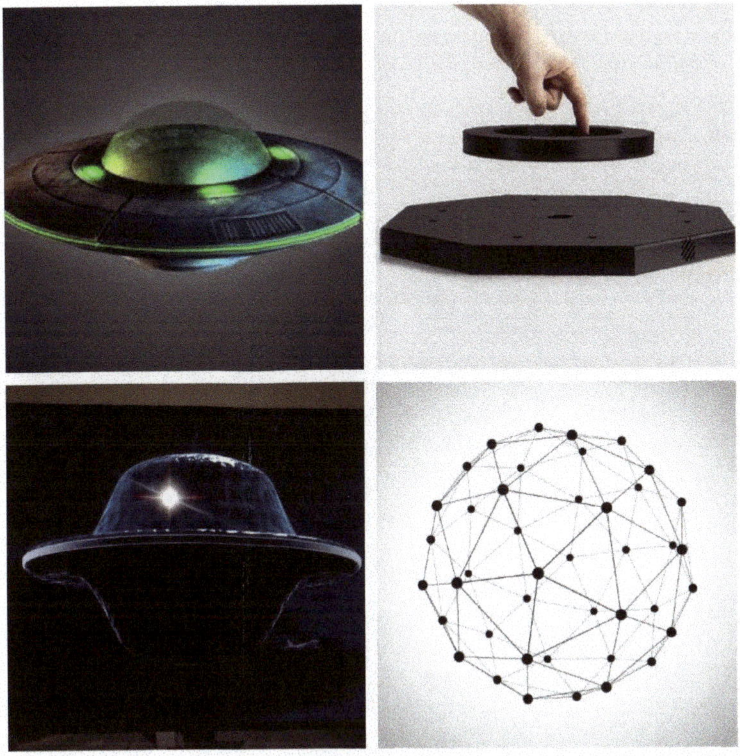

1. The Main Nuclear Engine

The core is located in the center and is the main nuclear power source of the apparatus.

2. Two Gigantic Magnetic Discs:

Both discs encircle the nuclear core, with one positioned in the north and the other in the south. The northern disc is the larger of the two. These discs, alongside other engine functions that contribute to the formation of the apparatus, play a role in the operation of Earth's electromagnetic field.

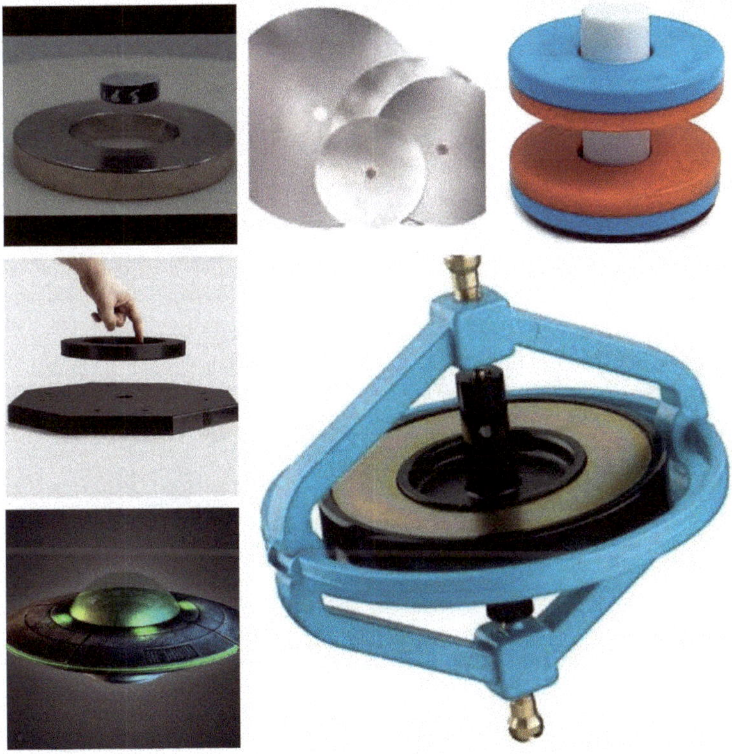

3. North and South Poles:

Two poles that operate as alternators or generators. One extending north and one extending south.

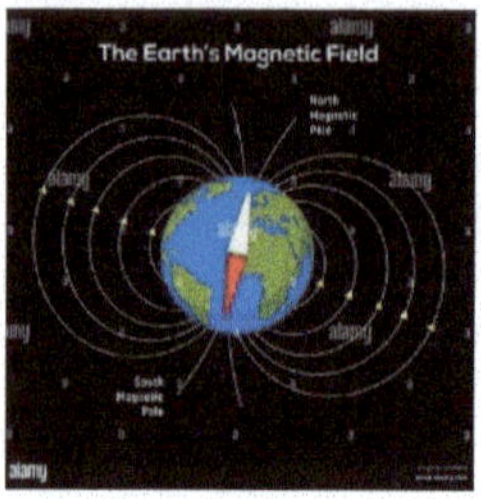

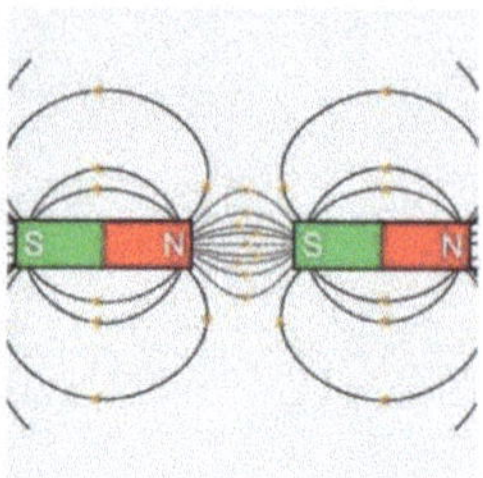

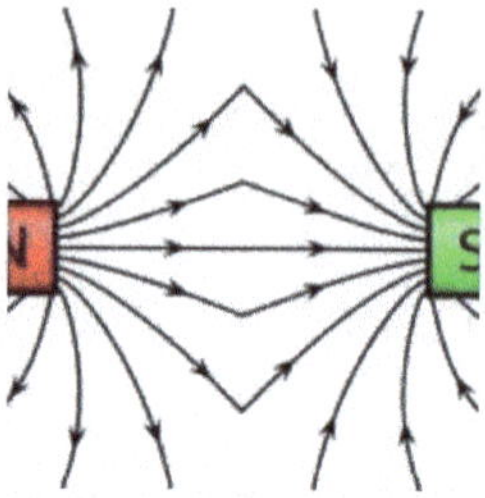

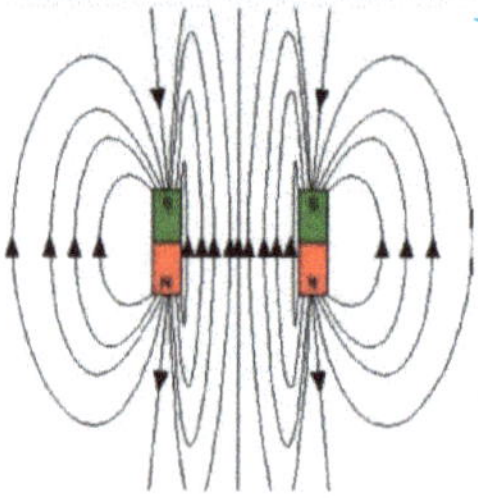

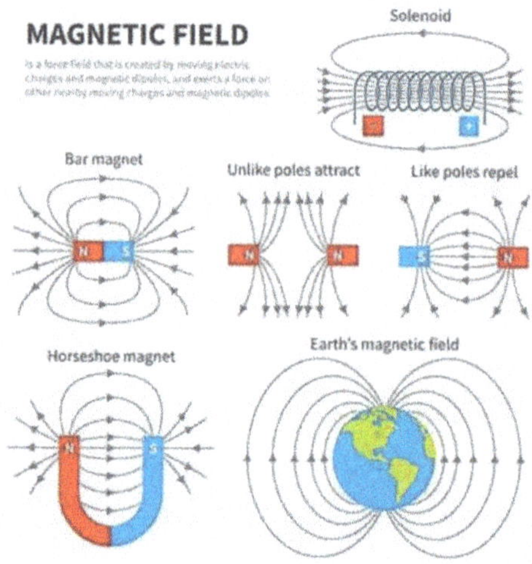

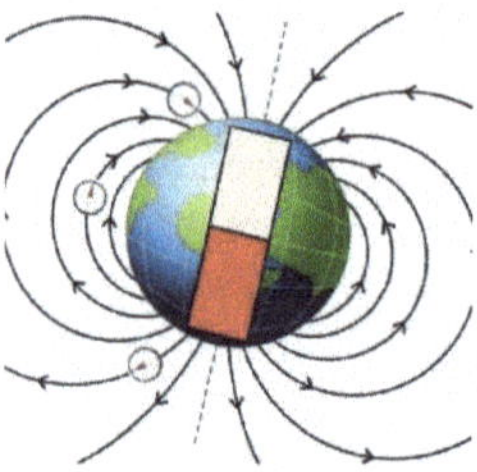

4. Turbines

There are seven nuclear turbines powered by the core nuclear engine. Four are positioned on top of the larger more powerful electromagnetic disc and operate separately from the northern magnetic disc, while three are situated on top of the smaller southern magnetic disc, and they also function separately.

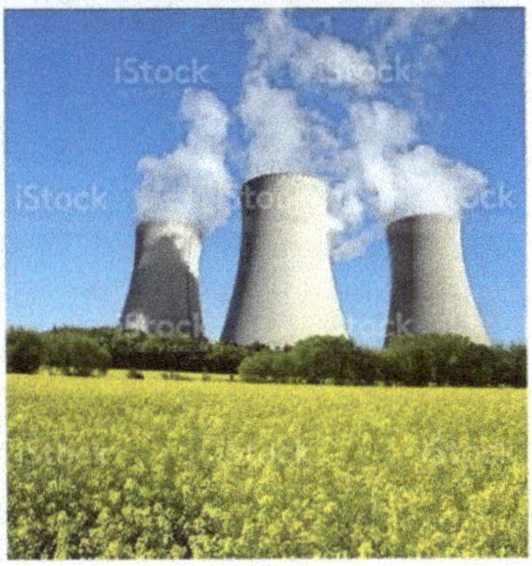

5. Transmissions and gears

These are connected to the nuclear turbines.

6. Drive shafts

These are in connection with the transmissions and gears.

7. **Pistons**

These are attached inside the drive shafts. They move the grinders up and down during operation. They eventually settle the grinders down into their winter storage slots.

8. The Grinders

These are seven sets of three iron balls that consist of spikes made out of diamonds. Four sets in the north, three sets in the south. Two out of the four sets in the north spin counterclockwise, and the other two sets spin clockwise. Two out of the three sets in the south spin clockwise, and the other set spins counterclockwise.

They pulverize almost anything. They are connected to the pistons, drive shafts, transmissions, and turbines.

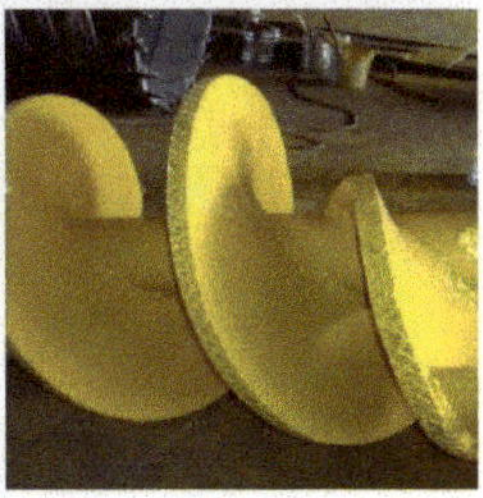

9. Double Loops

These are gold loops structured within the electromagnetic vortex cage. These loops provide reinforcement and sustain heat for the electromagnetic vortex cage, serving as conduits for electricity.

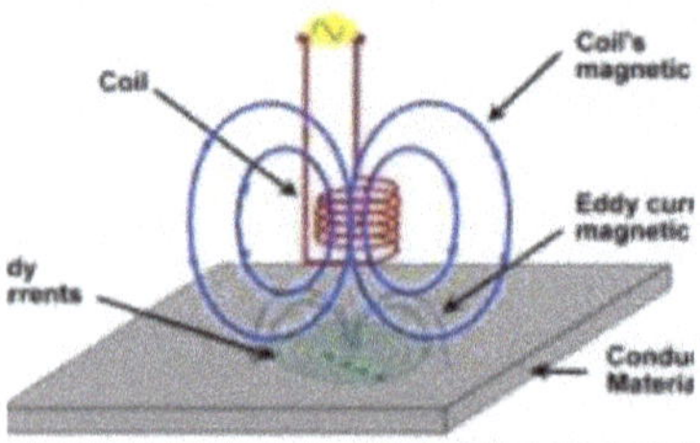

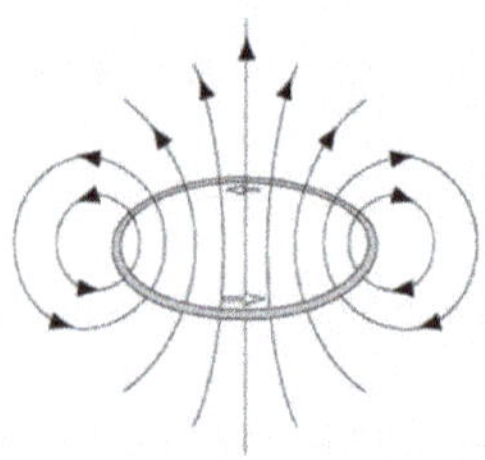

10. The Electromagnetic Vortex Cage

This acts as a containment and encapsulates the engines of the apparatus. The cage is composed of tetrahedrons connected at their points to form a smooth sphere. Moreover, the cage serves as a conduit for electricity.

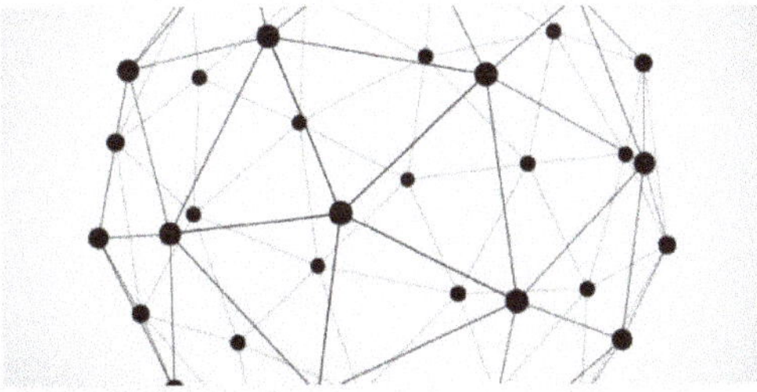

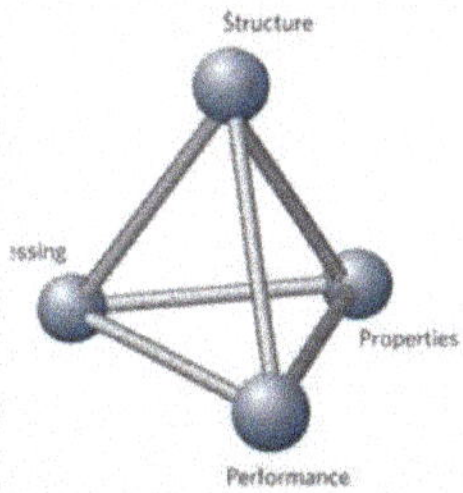

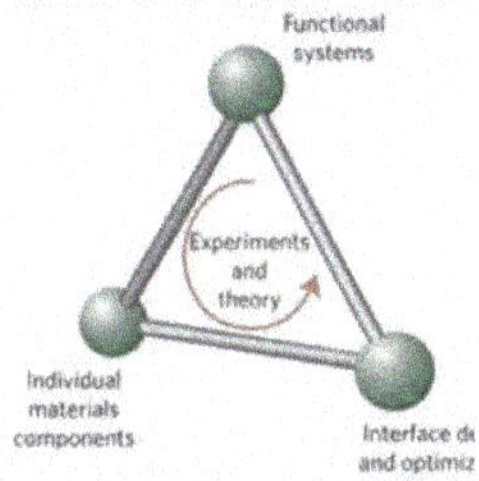

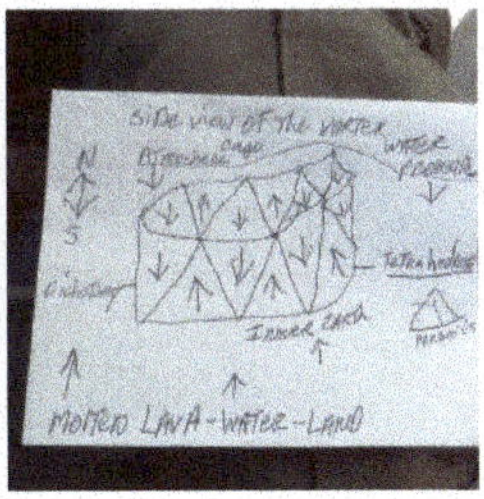

11. Water

Water acts like a pressurized cage, amongst many other functions.

12. Our Moon

It stabilizes our Earth from wobbling out of control. Anything else related to our Earth will be considered as effects. Let's start with a slight general knowledge of engines.

The main function of engines is to continuously produce power while maintaining stability and balance. In order to do these functions, an engine needs an overall equilibrium between staying cool and warm. The engine can't get too hot and can't get too cold. To start, there are four easy ways engines need to keep cool and warm, among many others.

1. Oil
2. Water
3. Wind
4. Exhaust

Let's start with water since the main engine is nuclear and water is the most abundant substance on Earth. Scientists agree that small amounts of radiation leak into the air from the surface of Earth. This diluted radiation is produced at a high rate from a very powerful encased nuclear engine located in the center of the apparatus, inside Earth. As the radiation from the central nuclear engine flows toward our Earth's surface, it loses potency and becomes harmless through filtration. Radiation is quickly soluble in water. This harmless diluted radiation is called 'Terrestrial radiation.' The reason why the results of studies on Terrestrial radiation show fluctuating measurements on our Earth's surface, instead of a precise thesis, is because of the different rates of dilution.

This is especially true when measurements are challenging to obtain from beneath massive bodies of mountains and water. These sweeping, fluctuating results are relatively easy to explain.

The engines of the apparatus are wobbling and spinning; therefore, the results of studies fluctuate. Again, this encased nuclear engine creates massive amounts of power, producing extreme heat and radiation. This is what scientists consider the core.

One of the functions of water is to keep things cool, and another is to act as a filter and solvent for radiation. For instance, removing a dry sticky spot on your countertop becomes easier to clean by just adding a little water. When someone gets overexposed to radiation, the first immediate treatment is to remove all your clothes, shower with warm water, and scrub with soap. Therefore, water dilutes radiation.

Water also functions as a pressurized liquid cage, encaging the Earth.

Going deep into the ocean without the correct equipment will end up in a catastrophe. The Earth is a series of complicated engines surrounded by entities that keep them cool, such as water and liquids. Obvious proof: Earth is around eighty percent liquids. Another easy observation is that instead of just agreeing that our Earth is spinning, we can easily observe and agree that **something** is making the Earth spin. The next engine, which encircles the nuclear core and is part of creating the electromagnetic field and gravity, consists of two gigantic magnetic discs connected together by opposite polar fields.

One magnetic disc, being slightly larger than the other, produces a dichotomy that creates balance and equilibrium.

The larger disc is the northern one, spinning counterclockwise, and the smaller disc is the southern one, spinning clockwise.

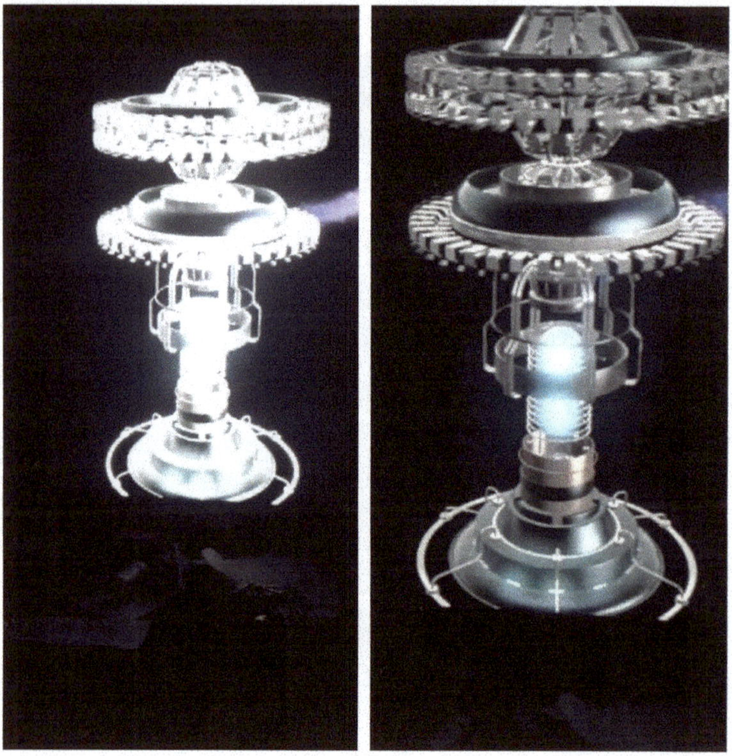

We have a scientific agreement that the whole Earth is encased in an electromagnetic field or shield.

Looking at the function of the largest, most powerful engine of the apparatus, the core, the center of our sphere, is the main engine that creates the nuclear power needed to spin the magnetic discs.

This part of the apparatus, which contains the magnetic discs, is the main structure that produces the electromagnetic field or shield. These forces cause the entire engine apparatus to spin counterclockwise, along with the entire Earth.

These major structures are combined and encased, operating as a unified whole. The forces generated are the result of cause and effect, action and counter-action, or reaction. Alternatively, it involves action, slowing down, followed by a balanced operation. The primary nuclear engine heats the core, generating power to spin the northern magnetic disc, the nuclear turbines, and the northern pole. The northern pole acts like an alternator and produces electricity. Nuclear power and electricity power the nuclear turbines, which operate the drive shaft's pistons and spin the grinders. The electricity also produces the power that electrifies the magnetic vortex cage. The golden double loops, which support the electromagnetic vortex cage, also act as a transformer, enhancing the flow of electricity around the electromagnetic vortex cage.

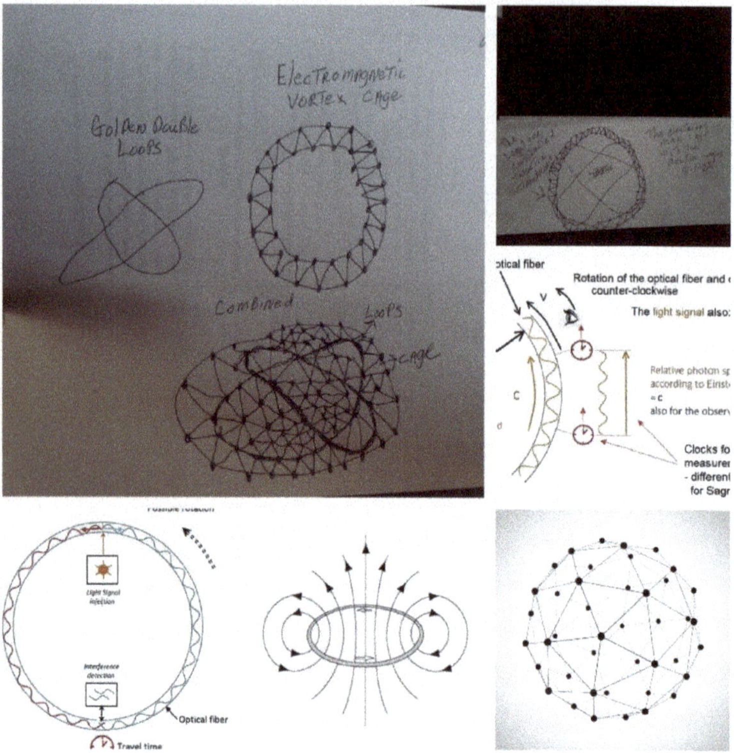

The double loop structure is very common in the field of electronics and the flow of electricity. The electromagnetic power pushing out to shape the magnetic field is also reintroduced into the apparatus, back into the nuclear engine. This process creates and shapes powerful loops constituting the electromagnetic field. The electromagnetic field's main function is to shield our Earth. Additionally, the nuclear engine produces power for the smaller Southern disc. This magnetic disc spins clockwise, powering the Southern pole, which also spins clockwise, producing electricity, such as an alternator or generator. The electromagnetic power produced by the southern pole (alternator) powers turbines, the drive shafts, and pistons, which spin the southern sets of grinders. The southern sets of grinders also guide electricity to the electromagnetic vortex cage. The golden double loops act as a conduit, producing the bottom half of Earth's electromagnetic field, or shield. This part of the apparatus induces the power of the electromagnetic shield back into the nuclear engine, which completes the double loop of the electromagnetic shield.

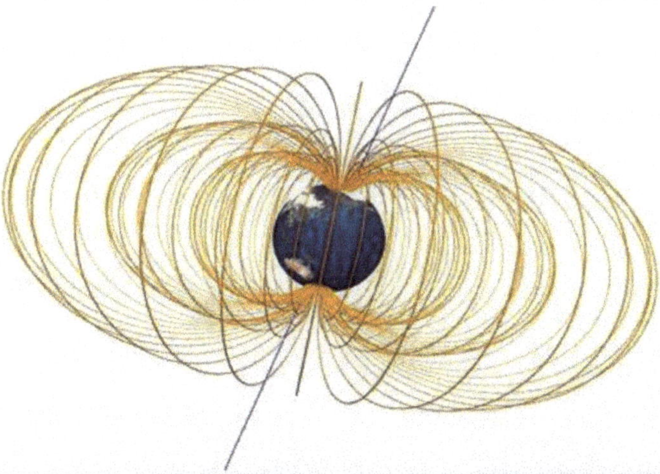

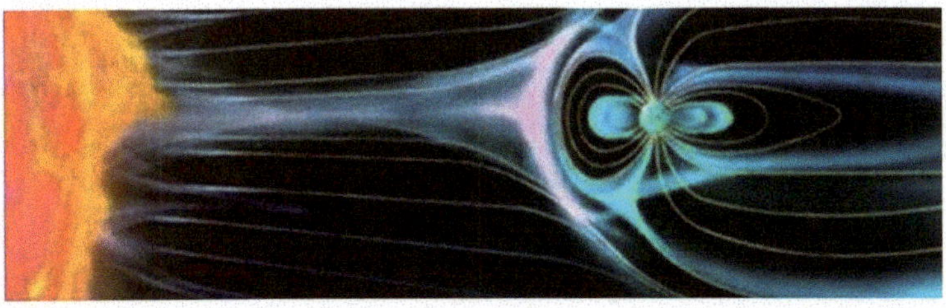

The whole middle section of the apparatus, which contains the main engines, including the magnetic discs and the nuclear engine, is entirely encased. These forces are creating dichotomies that work and balance each other. The main engine located in the center is nuclear fission. Take a smaller version of our Sun, place it in the center of the apparatus, and encase it in a solid sphere.

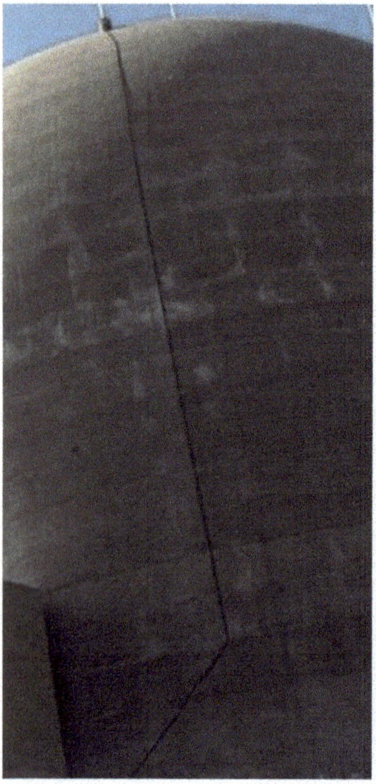

This is an example of a nuclear power plant, which is considered the core of our Earth by the scientific community.

Connected to this solid core are a north pole and a south pole. Both poles are part of a system that produces electricity, the same as alternators or generators. Again, a nuclear engine in the center of two magnetic discs powers the engines of the apparatus. Moreover, connected to the nuclear engine are two poles, one south and one north. The nuclear engine powers turbines, which spin both poles. There are other turbines connected to transmissions, then to the drive shafts that operate the pistons and spin the sets of grinders. The pistons move the sets of grinders in and out, starting from the end of winter until the beginning of fall. This up-and-down function happens in the north section and the south section of the apparatus, respectively. They rise during spring, then slowly shelter back down in the fall. Continuing down until they reach their winter shafts. The transmissions and gears control the drive shafts and the sets of grinders, which, in turn, pulverize the crust of the Earth.

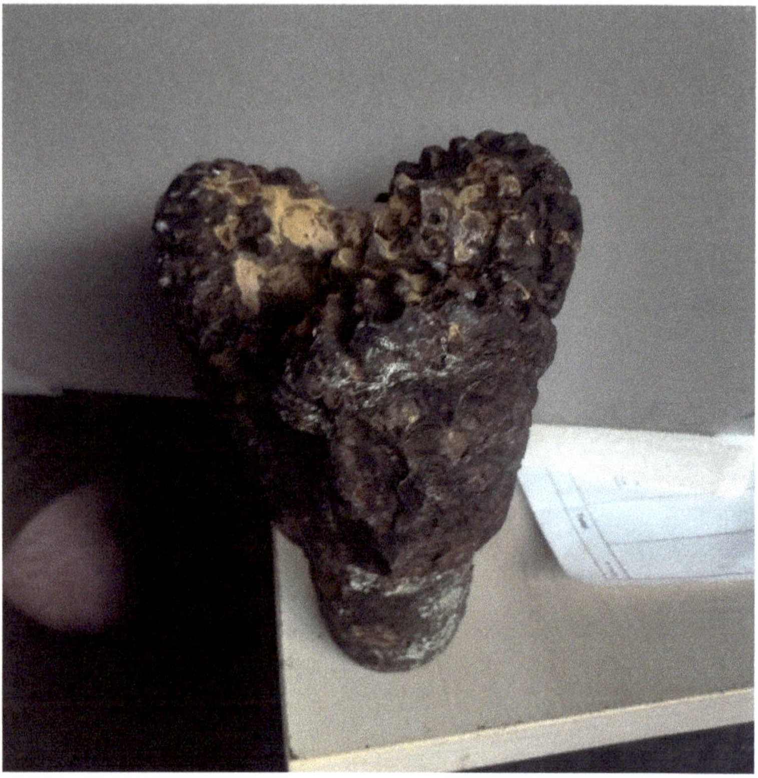

Let's go over this gradient again. Connected to the extremely powerful core engine are two magnetic discs. The northern pole (alternator) of the apparatus representing the positive side is connected to a transmission system equipped with gears. This system is connected to the top- center of the nuclear core engine turbines and spins counterclockwise. On the other hand, the southern pole (alternator), representing the negative side, is connected to a transmission system equipped with gears. This system is connected to the bottom-center of the nuclear core and spins clockwise.

These opposites, which are dichotomies, produce an extreme amount of electromagnetic energy. The main formation of the center engine becomes complete, spinning the entire apparatus and the Earth counter-clockwise. Extreme radiation, heat, and an abundance of magnetic force are unquestionably created. To give you a reality of our surroundings: heat, steam, lava, and ash, giving rise to volcanoes, the redundancy of weather patterns, fog, clouds, and other occurrences like the "Old Faithful" geyser in Yellowstone National Park, including our atmosphere is created by the apparatus.

The following are the four main necessities for a stabilized and balanced operation of an engine:

1. The oil keeps the whole apparatus cool and lubricated, which then provides a smooth operation of the apparatus.
2. Water keeps the apparatus cool and pressurized and filters radiation.
3. Exhaust is the steam, lava, ash, gas, and many other elements that extrude from the surface of the Earth.
4. The wind keeps the surface of the Earth cool. These are considered jet streams.

As many times as we go over a gradient, the more we become certain. Once again, two poles, one in the north and one in the South, serve as generators, also known as alternators. They are connected inside their drive shafts, which are connected to nuclear turbines. Altogether, both poles produce electromagnetic energy. They supply electricity to both the electromagnetic vortex cage and the electromagnetic field.

The northern section of the apparatus consists of four sets of gyros, or grinders. These four grinders, as a whole, are spinning counterclockwise since they are all connected to the northern disc. Of these four sets of electromagnetic spheres or grinders, one set spins counterclockwise. The next set in succession spins clockwise, followed by the next set in succession spinning counterclockwise, and the next subsequent in the sequence spinning clockwise. This pattern continues for a total of four sets. This function produces a continuous dichotomy while the pistons slide them up and down, one after the other.

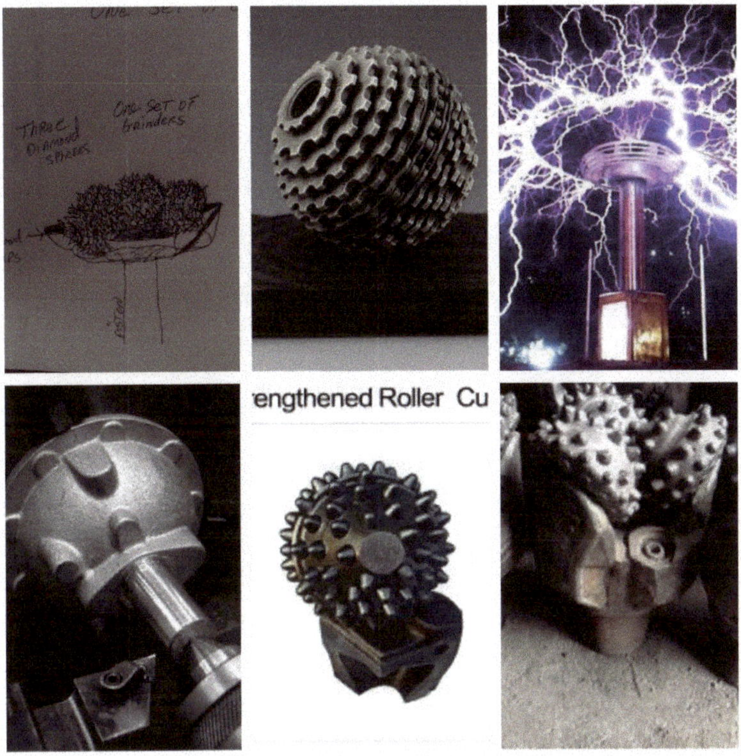

The main function of the grinders is to pulverize anything and everything in their path. Another function is to spin the Earth back and forth on its axis.

The northern section of the apparatus opposes the smaller southern section of the apparatus, creating a balanced dichotomy. The southern section of the apparatus produces a negative charge and has three sets of grinders, which overall spin clockwise since they are connected to the southern section. One set of grinders spins clockwise, the second set of grinders spins counterclockwise, and the last set of grinders spins clockwise.

The balance of cause and effect takes place because the northern section has more power than the southern section. The northern four grinders, which are part of the northern pole section, overall spin counterclockwise. The three southern grinders, which are part of the south pole section, oppose the north pole section by spinning clockwise. The north pole section and south pole section create a balance. The southern three sets of grinders keep the northern four sets of grinders from spinning and wobbling out- of- control. Both the north and south grinders create a balance by producing a braking effect. The engines of Earth have many separate functions, much like our Moon, water, our Sun, and foliage. The counterclockwise spin of the northern sets of grinders, balanced by the clockwise spin of the southern sets of grinders, also has other functions. Another function of this opposition is the tilting of the Earth on its axis, creating a hot and cold dichotomy by tilting toward and away from our Sun.

These two sections of the apparatus, north and south, also create a balanced dichotomy with the positive flow of the north spinning counterclockwise against the negative flow of the magnetic disc and the smaller southern section of the apparatus, which spins clockwise.

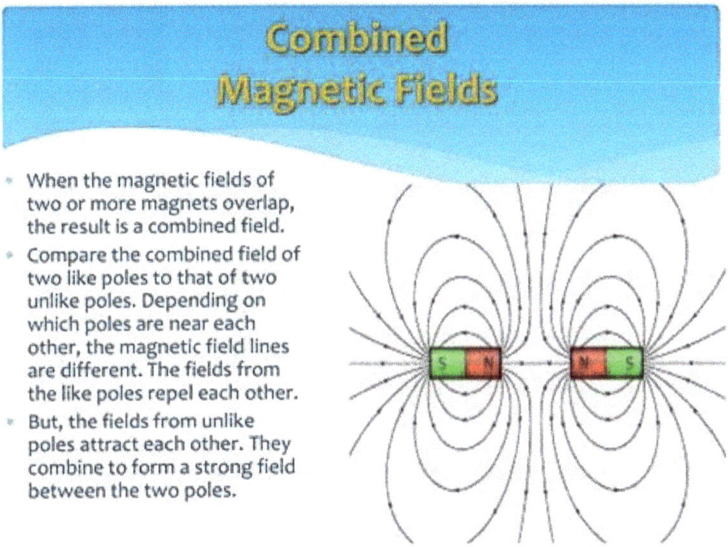

Just a reminder that, all in all, the northern part of the apparatus, along with the power of the center nuclear engine, spins Earth counterclockwise. The section of the southern apparatus spins independently clockwise, using nuclear power from the core.

This counter-opposition causes a braking effect and balance. The scientific views from space agree with these functions. Views from space, pointing cameras directly down at the southern pole, show video of the appearance of Earth spinning clockwise. The views from cameras pointing directly down at the northern pole show video of the Earth spinning counterclockwise. The scientific community agrees with these videos. Only during the fall and spring seasons do the four northern sets of grinders and the three southern sets of grinders spin at the same time.

These countering motions are the cause of the hurricane season in the fall and the tornado season during the spring. There are chapters dedicated to these effects, which go into more detail later on in the book.

Earth tilting back and forth on its axis is another scientific agreement. Looking at this agreement, we will use the winter and the summer solstices as examples. This agreement proves the existence of **something** creating this tilting effect. The cause is the counterclockwise forces being produced by the spin of the northern sets of grinders and the opposite clockwise spin being produced by the southern sets of grinders. The two powerful forces of opposition create many catastrophic weather phenomena.

Turning on, getting hot, slowing down, and cooling off in the north, followed by turning on, getting hot, slowing down, and cooling off in the south, create this magnificent balance. Yet, these massive forces cause catastrophic destruction. Although these causes are livable and easily predictable, we first have to agree that Earth is an engine in order for future predictions to take place.

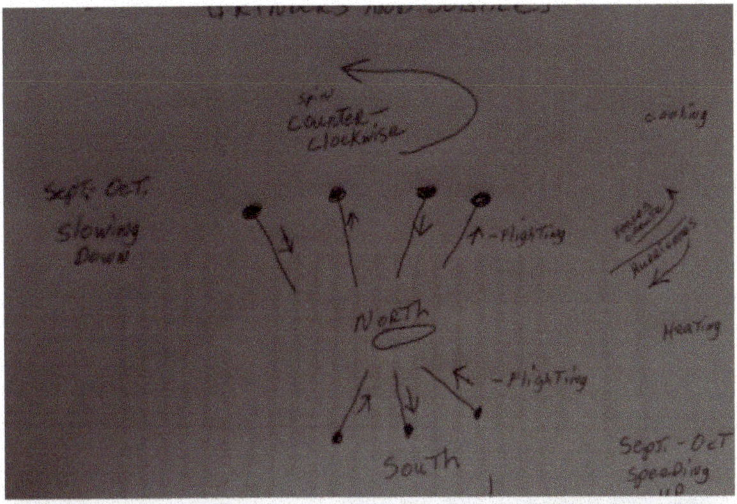

An example of proving Earth is an Engine is easy. First, we can agree that at the end of summer and the beginning of fall, something turns off and begins to cool down. Therefore, go get in your vehicle, drive it for awhile, turn the engine off, and then put your hand on the radiator or the exhaust pipe. Please, don’t do that, you will burn your hand.

For now, it doesn’t matter what you want to call the thing that has shut down inside of Earth as long as we can agree that something has turned off. This something has been running nonstop for around three months. This we call summer.

Then, **SOMETHING** needs around three months to cool down, which we call fall. In the past, this cooling down and shutting off was considered to be the result of Earth tilting on its axis. Now, we can easily observe that the tilting of our Earth is not the only function of the axis.

The axis also spins in connection to two poles, centered at the core with positive energy in the north and negative energy in the south, ultimately assisting the inner core of the apparatus with electrical power. This creates the power needed for Earth's electromagnetic field. Along the electromagnetic vortex cage (discussed in a later chapter), the northern and southern sets of grinders also use the energy being produced by the alternators or north and south poles. We just went over what is considered a steep gradient.

Let's go back and look at the different types of Engines we have on Earth. We have locomotives and tractors equipped with huge diesel engines. There are different types of jet engines powered by turbines. Additionally, there are immense engines in aircraft carriers, some with huge turbines using nuclear power. Nuclear submarine engines also fall within this category.

Nuclear engines produce massive amounts of power, spinning extremely large turbines with very little input. Yet, they also produce a massive amount of destructive radiation.

When we look at the different shapes and the shining brilliance of the stars, we can easily agree they are different. Astronomers even have names (agreements) for their different shapes and activities. Examples include quasars, pulsars, super giants, novae, and red giants. These are bright, shining stars, some with two points while some with seven. Black holes are considered massive gravity planets. None of these agreements (stars) are considered in the scientific realm as engines of different types.

This is simple. The stars are different types of engines. Planets within our solar system should also be considered different types of engines. Our Sun functions as an immensely potent nuclear fission engine. Jupiter, the colossal gas giant, is classified as a multi-purpose gas engine. Planets possessing rings, like Saturn, should also be classified as engines. Those with rings and moons equally qualify as engines.

Let's do away with this false concept of 'Mother Nature,' which is just an excuse for the unknown. Mother Nature makes the world go round? Not anymore!

Now, let's look at the engines of our Earth and move along on this gradient. The main engine of our Earth, and the largest, is a nuclear sphere located in the middle of the globe.

This engine of the apparatus does not spin along the imaginary line created by scientists called the equator. The spin of the entire apparatus tilts back and forth on an axis that is observed as a wobble. The spinning of the apparatus is not absolute; therefore, the overall route is on a variable that wobbles. The center engine is also made up of two gigantic magnetic discs held in place by the apparatus with their opposite charges facing each other.

Remember, an encased nuclear engine is located within a hole in the center of the two magnetic discs, producing a sphere in the center.

Consider a nuclear power plant placed inside the Earth, creating the core or center, which spins counterclockwise. The nuclear engine also provides power to seven turbines, with each turbine connected to the base of the four sets of grinders in the north and three sets of grinders in the south.

Each turbine is connected to a transmission with gears. The transmissions and gears control the drive shafts and pistons, which, in turn, spin the sets of grinders.

The entire engine apparatus includes the north and south poles, which are the Earth's axis and act as alternators or generators. The gold double loops, which support the inside of the electromagnetic vortex cage, also act as a conduit or transformers. The entire apparatus is being held in position and cooled by the combined heaviest entity on Earth, the pressure of water and liquids. Everything on Earth is being drawn, propelled, and pulverized toward the center of the apparatus. This is called centripetal force. Created by the spinning of the apparatus and using centripetal and electromagnetic force, this combined cause and function is considered throughout the scientific community as gravity. Therefore, the formula for Gravity is:

EMF + CF = G

Simple!

Electromagnetic Force plus Centripetal Force equals Gravity. Based on the equation above, we can see how to control gravity by the size, spin, and power of the magnetic discs. The faster the spin and the higher the production of electromagnetic force, the greater the centripetal force. Therefore, the power of gravity can be controlled.

The complex topic of oceans and gravity will be discussed in more detail later in the book. The nuclear power that spins the giant magnetic discs produces electromagnetic energy. Again, this electromagnetic energy, combined with the speed of centripetal force, is gravity. Two of the northern sets of grinders that spin independently clockwise; pulverize and push materials from the surface of Earth down into the center core. The material gets heated, electrified, and melted into what is considered lava. Then, this heated substance gets pushed back up by the other two northern sets of grinders, which spin counterclockwise. Through the upward spin of the flighting, the material gets pushed to the surface with centrifugal force, where it begins to cool. The centrifugal force is created by the counterclockwise spin of two out of the four sets of grinders. The lava gets cooled, and the liquid flow of the smaller portions gets filtered and electrified through the electromagnetic vortex cage. This centrifugal force and cooling process forms the Earth's crust, land, hills, and mountain ranges.

The larger chunks of metals and materials get caught up with centripetal force, pulverized, and propelled back downward by the other two grinders rotating clockwise to undergo processing once more until the substance is capable of flowing through the tetrahedrons of the electromagnetic vortex cage.

I have explained an overview. For many, the agreement that Earth is an engine needs more explanation. Let's start from the inside and work our way out again.

The Core

The core is an extremely powerful, encased nuclear engine in the center of two gigantic magnetic discs that are locked together in opposite polar fields. The discs are being held relatively apart. This opposition produces a dichotomy, with the larger northern disc spinning counterclockwise and the southern disc spinning clockwise, which in turn creates the electromagnetic field. This opposition of both discs spinning produces stability, equilibrium, and balance.

The Grinders

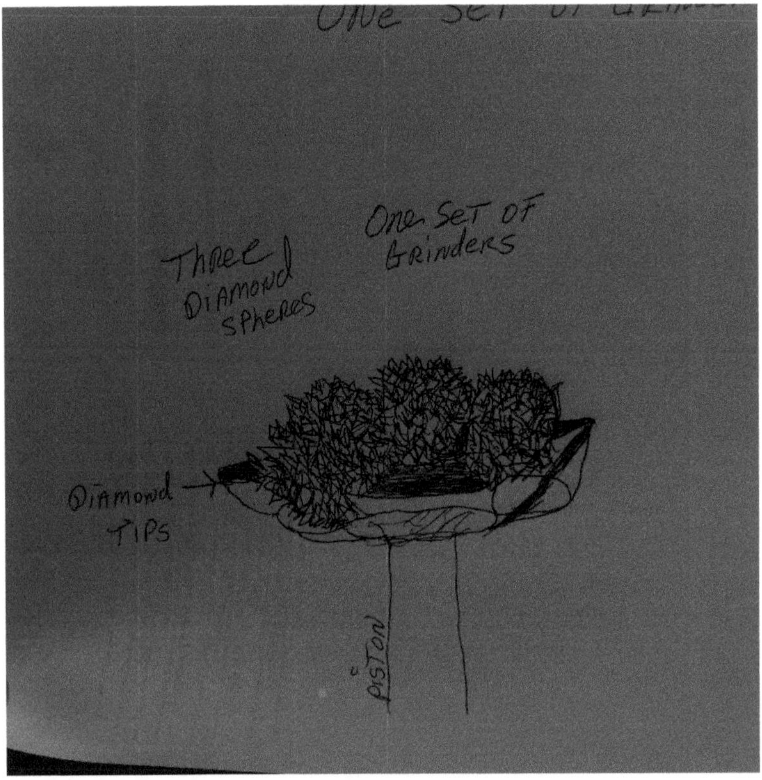

The northern and southern sections of the apparatus have sets of grinders. The sets of grinders are spheres connected to drive shafts with auger flighting blades. The grinders connected to the drive shafts have spiral metal waves along the length called flighting. The spiral flighting guides material, such as mud, clay, rocks, metals, liquids, etc., toward the center of the core using centripetal force. Also, the opposite sets of grinders have reverse flighting, rotating up and back toward the electromagnetic vortex cage. This flighting aligns with centrifugal force or inertia. The centripetal force and the electromagnetic force created by the main engines heat the materials of the Earth. Then, the materials get pulverized by the spheres of spiked diamonds into lava, gas, and many other substances. The inertia from the centrifugal force pushes the lava back up toward the sets of grinders. The flighting on two of the sets of grinders, which spin counterclockwise, push the material toward the crust of our Earth. On its way back up, the two sets of grinders pulverize the material again and push it out toward the surface. A good example is a regular household screw. Look at the grooves on a screw; when using a screwdriver, and you turn the screw to the right or clockwise, the screw goes into the surface. When taking out a screw using a screwdriver, and you turn the screwdriver to the left or counterclockwise, the screw is removed from the surface. As the saying goes, "Righty tighty, lefty loosey." The lava cools and is forced out through the tetrahedrons. The tetrahedrons create the structure of the electromagnetic vortex cage. As the lava cools, land and other entities are created. These grinders and drive shafts with pistons create an up-and-down track that can clearly be seen on weather maps.

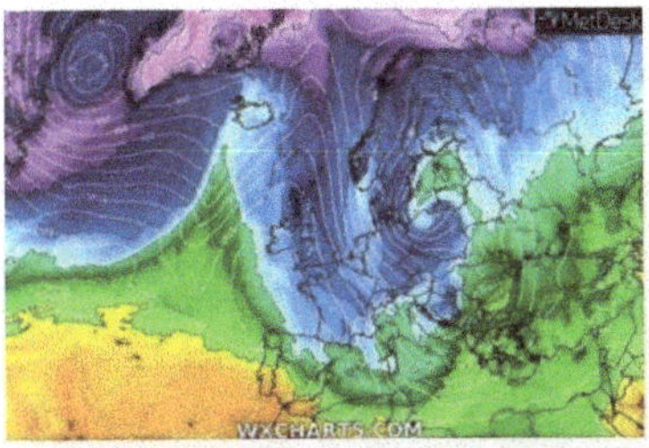

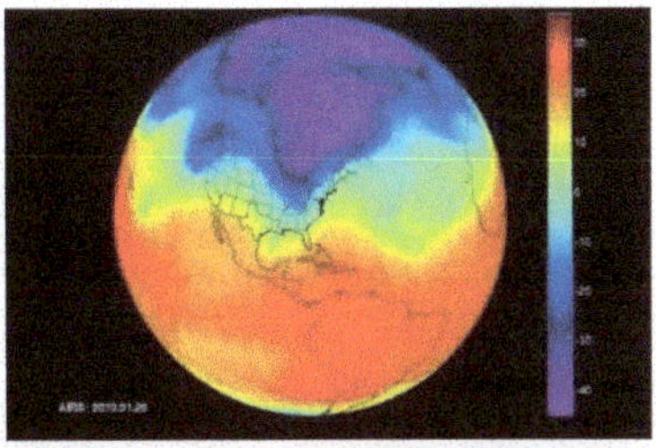

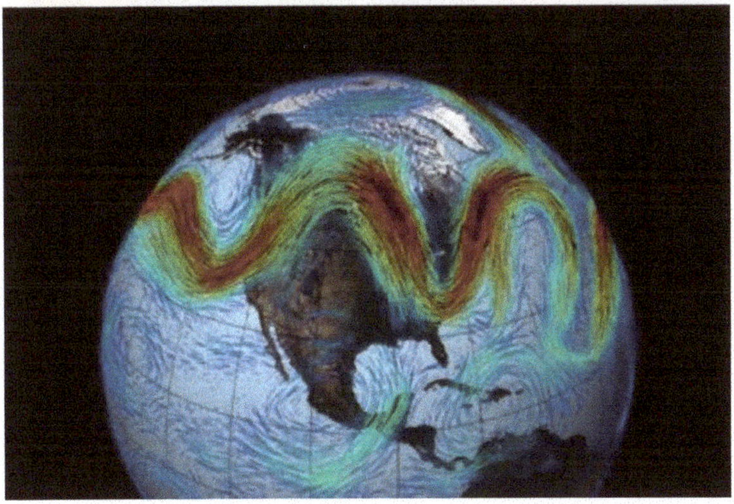

They begin to warm up and cool down at the same time, respectively, during spring and fall. The northern sets of grinders and the center engine act as one and spin the whole Earth counterclockwise. The weaker southern section of the apparatus spins clockwise, creating balance and equilibrium by creating a counter-action or braking effect (dichotomy).

From the positive North Pole to the negative South Pole, both serve as electrical conductors, generating electromagnetic energy through the same principles used by alternators and generators. The electromagnetic energy produced by the rotation of the magnetic discs and the spinning of both poles and the nuclear core completes the apparatus. These conductors of electricity form the basic agreements of generator technology. Also, the grinders produce electricity on a microwave simplicity, causing humidity.

I'm not claiming that I can design the engines of our Earth or that I'm one hundred percent correct. I have absolutely no education in technology, physics, engineering, meteorology, or astronomy. This book is only the beginning of realizing how Earth actually exists.

The northern sets of grinders slowly stop spinning and slide back down into their winter position around the end of September, completely stopping around the winter solstice. In the south, simultaneously around the end of September, the grinders slide out of their winter position and begin to warm up.

During the end of winter, the northern four sets of grinders, around the end of February, begin to move out of their winter slots and begin to warm up, then two sets slowly start to spin counterclockwise, and two sets spin clockwise faster and faster. This agreement is called the beginning of spring in the northern section of Earth. At the same time, the southern sets of grinders slowly begin to shut down. The three sets of southern grinders gradually begin to slow down and move back into their winter slots. This agreement is called Fall in the southern section of Earth.

The gradual slowing of rotation in the clockwise direction in the south and the initiation of counterclockwise spin in the north results in the tilting of our Earth on its axis. The springtime in the north continues to get hotter, and at the same time, fall in the south is cooling, and their three sets of grinders return to their winter slots or positions. When we get to the northern summer solstice, their four sets of grinders begin to slow their counterclockwise spin, close back into their winter positions, and eventually cool, resulting in the fall season. When we get to the southern winter solstice, at around the same time as the northern summer solstice, the southern sets of grinders slowly begin to move out of their winter positions and warm up.

Fun fact: The academic community study only two solstices. Observations reveal four, two in the north and two in the south. An easy explanation for the academic community having to study four solstices is that they are all different from each other. The solstices are effects resulting from the sets of grinders turning on and shutting down. This happens four times, two in the north and two in the south. The north has a winter and a summer solstice. The south has a winter and a summer solstice. That makes four.

The result of these two opposite forces creates the tilting of the Earth on its axis. Also, it creates a dichotomy. Again, dichotomies create balance and equilibrium. There are four major entities on Earth that are designed to keep the engines cool. Extreme heat and extreme cold are the major causes of engine breakdown or engine failure. Every engine needs something to keep it stable and cool. Again, the four major entities or components of keeping engines cool are oil or lubricants, water, wind, and exhaust.

Stabilization

Engines of all types need to be stabilized. An engine that is not stabilized spins out of control. Observations of our solar system show us how the planets revolve around our Sun.

Just like any engine, in order for them to operate, they need to be bolted and mounted. Obviously, the planets are not bolted or mounted to anything, yet they are being stabilized. One of the main stabilizers of our Earth and many other planets is their moons. Our Moon stabilizes the wobbling and positioning of our Earth by pulling the water and liquids from one way to another. We agree to call this reaction 'Tides.' More on this subject will be explored in the chapter titled, 'Our Moon.'

Recapping

Our Earth overall is spinning counterclockwise. Through basic observation, at the end of February and the beginning of March, the cycle of the northern section of the apparatus begins, which consists of four sets of grinders tucked away for the winter. The grinders of the north with flighting, such as augers (attached wavy blades) that run along the outside of the drive shafts, begin to spin. As they begin to warm, open up, and slowly spin counterclockwise, they pick up speed. Two sets of the diamond-spiked spheres begin to grind by spinning clockwise, and the other two begin to grind counterclockwise. Remember, the whole apparatus continues to rotate counterclockwise.

In the northern lands, we call this warming season spring. Around the same time, the southern engines and three sets of grinders begin to gradually cool and slow their spinning, moving back into their winter position.

This is called the southern fall. Around the beginning of March, the northern sets of grinders open on a gradient and slowly spin, building up speed and causing strong winds. The electricity from the grinders and electromagnetic vortex cage produces the effect of a beautiful, cool, Sunny March day. Also, in March, starts the beginning of different weather patterns with cumulonimbus, or lightning clouds. These clouds have been absent in the skies since the shutdown of the northern sets of grinders at the end of September. While electricity and microwaves shoot up through the electromagnetic vortex cage, we can observe this effect by knowing that **something** is causing these cloud formations filled with electricity. When a grinder passes by and creates centrifugal force, the electricity and microwaves create a bright, beautiful day. This is called a 'high" in meteorology.

When the grinders speed up, the storms get stronger, and so do the hot, Sunny, blistering days of the summer. In the north, this pattern persists until approximately the end of September, at which point the weather patterns undergo another shift. Although all clouds are effects, cumulonimbus clouds only form at the end of winter until the end of summer, the beginning of fall. This sequence begins approximately around the end of February and slows down around the end of September. From the end of September until the end of February, in the northern section of Earth, cumulonimbus clouds and lightning storms won't exist. The exception is that, throughout the rotation of the central portion of the apparatus, considered the real equator, storms with electricity happen all year- round. The main cause of lightning and thunderstorms is the electrical power of the grinders. The accumulated electricity forms cumulonimbus clouds. Every year, give or take a few days, the northern grinders continue to open and speed up until they peak, which is considered the summer solstice. Then, the northern engine begins to cool and slow until around the middle of September. Let's slow down on this gradient here and ask a question: "How do we know?" March has a famous saying, "In like a lion, out like a lamb." While during other parts of the year, we hear people say - in April, "It's warming up." And in September, we hear, "It's cooling down."

Observations prove that in the north, **SOMETHING is WARMING UP!!!** In the south, **SOMETHING is COOLING DOWN!!!**

We reached a point of no return. We reached a gradient that is going to change the world. The next chapters and the rest of the book describe the effects of these engines. The effects further prove, through observations, that **something** is making the world go around. Going back over this chapter is something we all should study once we agree that **Earth is an Engine.**

Reality through observations makes this simple and proves they exist. Let's go over this again, step by step, starting with a little violence, **'The Grinders' or 'The Pulverizers.'**

Chapter 8: Gyroscopes and The Grinders

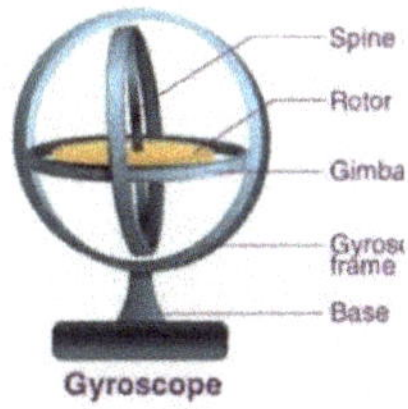

Strengthened Roller Cutter

Experiments with gyroscopes show the defiance of gravity. The results of these experiments are close to the design of the engines of Earth. Obviously, the experiments leave out a counterbalance - a counteraction that slows the gyroscope structure down, ultimately creating an equilibrium. The experiments leave out the dichotomy. The northern section and the complete apparatus spin counterclockwise; proof of this statement has been observed from space looking straight down at the North Pole. Also, looking straight down at the South Pole, the observations from space show that the Earth appears to be spinning clockwise. Overall, in general, the Earth is spinning counterclockwise. Observations of this is true by our Sun appearing to rise in the East and set in the West. This overall counterclockwise spin is produced by the northern sets of grinders and the main engine. The northern sets of grinders are located above the core engine, which is located in the center of the system.

Let's start with one gradient, increment by increment, step by step. Remember that the northern observations of Earth from space show the northern and southern sections spinning in opposite directions. This proves a dichotomy is taking place. An opposed force that creates balance and equilibrium. Therefore, Earth has been proven to be a dichotomy, yet this reality has not been revealed or studied. Until now, this revelation was handed over to an unknown and uncertain concept called 'Mother Nature.'

After the northern section is explained, we will continue down to the southern section and then examine the inside engine, which is the main force. Then we will work our way out, nice and easy. Let's look at the northern and southern compartments of Earth, which are located at the top and bottom of the system or apparatus.

Starting in the north, there are four sets of spheres or grinders. These are connected to pistons and drive shafts. Along the outside of the drive shafts is auger flighting.

This flighting can be observed on augers. These shafts extend downward and link to transmissions and gears. These transmissions and gears are connected to nuclear turbines. The turbines receive their power from the main engine at the center or core of the apparatus. Overall, the northern part of the apparatus spins counterclockwise. There are four sets of grinders in the north. Each set of grinders contains flighting along the sides. As the apparatus wobbles, each set of grinders individually moves up and down.

This up-and-down movement, the spinning and flighting, drives elements from our Earth, such as rock, liquids, soil, and debris, down where it heats up. Then, the heated material gets pushed back toward the surface through the electromagnetic vortex cage, creating land.

Altogether, the grinders connected to the drive shaft sections of the apparatus are spinning counterclockwise. Each distinct set of grinders is rotating either in a clockwise or counterclockwise direction. Once more, in the northern region, there exist four sets of grinders. Each set consists of three diamond spiked- sphere- metal pulverizers.

The first set spins counterclockwise. A second set of grinders spin clockwise. A third set of grinders spins counterclockwise, and the fourth set turns clockwise, thereby completing the cycle. This is a continuous dichotomy, and as we learned earlier, dichotomies create balance and equilibrium.

The Earth's crust, comprising of debris, man-made waste, soil, trash, junk, dirt, rocks, metal, ore, and all the elements, undergoes grinding and is propelled downward by centripetal force into the core of the apparatus. We can easily call this central location 'The furnace.'

Just like a blender, everything on the surface of our Earth is spinning into the center of the apparatus through centripetal force. Then, the extreme nuclear heat reduces it into lava. The lighter material gets pushed out toward the surface of Earth through centrifugal force, where it cools. The heavier materials are once more subjected to being pulverized and pushed back down through flighting and centripetal force. The cycle repeats itself.

Later in the book, we will discuss the exhaust and surface formations of the heated-melted lava material. Then, when the northern grinders shut down and slide back into the drive shafts, back into their winter positions. The grinders slowly come to a stop, in their winter position, and cool. This is considered fall and winter. Even though the grinders shut down, the northern section of the apparatus continues to spin counterclockwise through the forces produced by the nuclear core engine spinning the electromagnetic discs.

At a slower pace, the southern grinders slowly turn on, open up and begin to spin. This is called spring and summer in the south. Then, the grinding process is the same. The southern engine only has three sets of grinders; therefore, there is less centripetal and centrifugal force. Proof of this counteraction is by observing less land mass in the south compared to the north. The combination of the northern sets of grinders and the southern sets of grinders creates a dichotomy. They create the balance needed to float, rotate, and revolve around our Sun. Again, the northern grinders spin counterclockwise, and the southern grinders spin clockwise, forming a dichotomy. These engines combine to create a good part of gravity.

Again, in the south, there are three sets of grinders; one set of grinders spins clockwise, and the second set of grinders spins counterclockwise. The third set of grinders spins clockwise.

Two sets of grinders drive the materials of our Earth down toward the furnace along the flighting. The other one drives the melted materials and gases back up, along the flighting, toward the surface through the tetrahedrons of the electromagnetic vortex cage. The cage will be discussed in more detail later in the book.

Continuing on with our journey, we need to discuss the Electromagnetic Field.

Although the field is the last entity protecting Earth before our Moon, it is being created from the center of the apparatus. Therefore, let's journey there first.

Chapter 9: Our Earth's Magnetic Field

Here is a proven entity that absolutely exists.

An agreement that has been around for a while. Through observations, we can easily prove how the magnetic field functions. When objects from space enter through the field, they burn into small pieces. Numerous studies have proved that the electromagnetic field shields our Earth from the harmful radiation of our Sun. We can all agree that the electromagnetic field functions as a shield. When meteorites and debris from space fall through the electromagnetic field, they become hot from friction and burn up into harmless material (harmless to our Earth). However, the material could cause damage to the population and structures we have created, although, damage to our Earth, and damage to the engines that remain well protected. The electromagnetic field exists, overall, for the protection of our Earth. This protection provides a place where existence on the surface will still have a place to survive. Therefore, the electromagnetic field's number one priority is to protect our planet and the apparatus. This is proven by the damage a relatively small projectile from space can cause. Yet, little to no damage will be caused to the systems of the apparatus. A meteor can ignite acres of forest, mostly killing everything in its path, including humans. Eventually, the forest will grow, and societies will restructure. Also, just like all entities, we have to examine their functions, and overall, the function of the magnetic field is to protect the apparatus. Overall, the electromagnetic field acts as a shield. A good example is the body of a vehicle. The electromagnetic field is a shield and is the final element of our Earth. The magnetic field is the last filament of the systems of Earth's engines.

Have there been any instances where the magnetic field's protective function was compromised or affected by natural events or human activities? The answer is no, not permanently. Consider this: the electromagnetic field is like a product of a well-functioning machine, an effect of our Earth's engines. To disrupt it, we would have to damage the machine itself, like removing the essential oil and replacing it with a less effective coolant.

Even if we were to blast a hole in the magnetic field, it has a remarkable ability to replenish itself as long as the engines that generate it remain strong and intact. The electromagnetic field serves as a resilient guardian, ensuring the preservation of our planet's delicate balance and the protection of life on its surface. When we ask, "What entities are used to produce the functions of the electrical magnetic field?". The answer is easy: electricity and magnets. This is constructed in the center of the apparatus, at the core of our Earth. The answer is magnets and electricity.

Can the fluctuations in Earth's electromagnetic field affect climate or weather patterns?

Fluctuations in Earth's magnetic field have minimal impact on climate or weather patterns. While the electromagnetic field plays a crucial role in shielding our planet from harmful cosmic radiation, meteorites, and solar winds, its fluctuations are typically gradual and subtle. These variations do not exert a significant direct influence on the Earth's climate system. The electrical magnetic field is created by the apparatus and the engines of our Earth. Protect these entities, and the Electromagnetic Field will protect our Earth. Next, let's figure out forces most of us never learned, only experienced.

Chapter 10: Centripetal and Centrifugal Force

Centripetal and centrifugal forces function as a dichotomy. Centripetal force drives entities toward the center of a spin, while centrifugal force pushes lighter material toward the outside of a spin through inertia.

An Electromagnetic field flows toward a common center of the Earth. This common center is the exact location of the apparatus.

The more this electromagnetic force spins into the center, the more the energy becomes condensed.

I'm not a scientist.

I'm not an engineer.

I'm not an astronomer.

I'm not a meteorologist.

I'm a simple man with simple observations. Also, when we look at a blender and the blender spins, we see a hole in the middle. This is because heavier ingredients are drawn to the center by centripetal force, while lighter ingredients are pushed out along the outer part, called centrifugal force or inertia. If the sides of the blender had holes, smaller particles would have passed through the holes. However, the larger material gets caught back up into the spin, and centripetal force drives them back toward the center to get grinded again.

The same is happening inside our Earth. Even though the process is not a blender, the centripetal, centrifugal, and inertia forces are the same.

When lava reaches the electromagnetic vortex cage, instead of passing through holes, the lava passes through tetrahedrons. The tetrahedrons are the structures that make up the electromagnetic vortex cage; they function as funnels. The lava particles are hot while passing through the tetrahedrons; subsequently, they commence cooling, and as this cooling progresses, solidification occurs. This cooling forms Earth's crust and the land masses we see on the surface of Earth.

Indeed, specific regions on Earth experience more pronounced effects of centripetal, centrifugal, and inertia forces. If you follow the path of the grinders, you will discover that these paths differ from the course taken by Earth's most potent nuclear engine at the center core.

The center nuclear engine has the most pronounced influence of gravity. The trajectory occurs along the path of the center engine, or the core of our planet. We can now refer to this location as the real equator. In this critical region, called the real equator, the centripetal, centrifugal, and inertia forces are more pronounced due to the consistency of Earth's rotation. The pronounced forces of this region create the most fascinating effects on the planet, yet they also can be very devastating.

Some of the devastating effects caused by these forces are hurricanes and Earthquakes.

Understanding these unique regions sheds light on the intricate interplay of forces that shape our world and the delicate balance that keeps it in motion.

Centripetal, centrifugal, and inertia forces play an important role in the motion and stability of extreme celestial bodies in our solar system.

Therefore, let's figure out how *gravity* works!

Chapter 11: Gravity

The first thing I would like to start with is that gravity is circular on planet Earth. The apparatus, which contains the Engines of Earth, produces circular gravity.

Just look at the trees; their overall appearance is round. Their trunks and branches are round. The stems of plants are round. When flowers bloom, they open into a circle.

Bugs have round sections. Humans are round.

Our heads, arms, legs, toes, fingers, and butts, all exhibit roundness. I mentioned to a highly educated person how trees, overall, are round, and his answer was, "The Sun rises along the sky as the tree grows over the years. This growth follows the rays of the Sun, making the growth of the tree round." Also, he mentioned when the wind passes through the tree, the circular growth of the tree allows the wind to pass, making the tree stronger against nature.

My rebuttal was, "Even when trees grow toward the light, they still grow to be circular." I showed him trees that grow under a bridge, only getting half of the Sunlight from one p.m. until sunset. The bridge protected the tree from the wind. The same type of tree, and it was still round. If his answer was true, there would only be half a tree.

Also, you can place a tree in a sunroom with artificial light, and it will still grow to be round. This is not the only proof we can observe and study. Animals of all kinds, of all species, are round.

Take a bottle of any kind of sticky liquid found in your refrigerator or cabinet. Squeeze or pour one drop onto the floor. What do you see? When the drop hits the floor, what do you see? A round, circular splat.

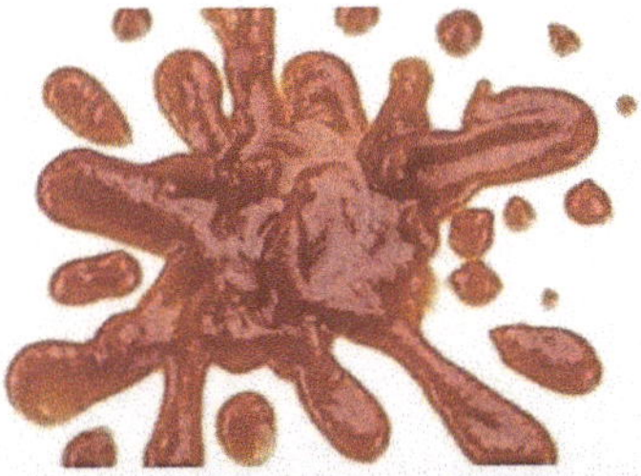

This is proof that gravity is circular and round. How do we know? Because we are spinning here on Earth, and we are on a sphere. It's that simple. Can we all agree on the statement above? As the globe spins, it wobbles. Then, of course, gravity spins and wobbles, causing the Earth, as a whole, to do the same.

Noticing mesmerizing pictures of decay, like those captured in the beautiful 'Arches' of the National Park of southern Utah, also shows proof of how gravity operates. Beautiful pictures of arches and caves along coastlines of the world also show the circular decay of gravity.

Amidst this captivating scenery, we find evidence of decay being caused by the circular presence of gravity. From cells to the grandest landscapes of our Earth and celestial bodies, we can clearly observe that gravity is circular and round.

This explains certain observations of the fluctuations coming from the electromagnetic field because gravity is not constant. Gravity fluctuates!

This fluctuation is the answer to planes mysteriously disappearing and the unknown sinking of ships all over Earth. Gravity is the strongest along the core engine, located in the center of the apparatus inside the globe. The core where the gravity is strongest should not be compared to the man-made, make-believe line- called 'The Equator.'

The apparatus inside Earth also wobbles and fluctuates. The Earth's core functions as a nuclear engine with two magnetic discs connected by opposing polar fields. When this core section of the apparatus and magnetic discs spin, they create centripetal force. Centripetal force draws entities toward the center. The magnetic discs create extremely powerful magnetic forces, which also draw all metals toward the center of the apparatus.

This spinning also creates the other side of a dichotomy called centrifugal force. Centrifugal force has inertia, which pushes the lighter entities out toward the edge of circular motion.

Along with the electromagnetic force created by the magnetic discs, there is centripetal force. Together, they hold and stabilize the apparatus at the core.

Therefore, the equation for gravity is: EMF plus CF equals Gravity.

Earth is not constant; therefore, gravity is not constant.

We are floating in space. Therefore, our Earth is not that precise. Science has us studying fake lines that fluctuate, called latitude, longitude, and equator, as though they are stable. Since the Earth is wobbling, gravity wobbles. Gravity is not constant. This is due to gravity not being generated precisely from the center of the Earth, as the apparatus is wobbling. The instability stemming from centripetal, centrifugal, and electromagnetic forces results in fluctuations. Consequently, gravity is more potent in proximity to the central organization of the apparatus.

Fun fact: Since we are spinning and the spinning of Earth creates time, if we stop spinning, time will change.

The next step out toward our Sun is the cage of the apparatus.

Chapter 12: The Gold Double Loops

The double-loop system consists of two gold metal loops, which are constructed inside the electromagnetic vortex cage. They are part of a complete structure that maintains the configuration and stability of the entire apparatus. Just like many entities we talked about earlier, they have more than one function. They reinforce the electromagnetic vortex cage.

Being made out of gold, the loops function as conduits for electromagnetic energy. Besides water, gold is the best conduit for electricity. The reason why gold is superior to water is because gold is solid, whereas water is a liquid and is unstable. The double bands are encased with an unknown metal.

As the Moon's gravitational pull keeps the apparatus from wobbling out of control, it needs an electromagnetic path to follow- a counteraction. While the Moon follows this path, it exerts a pull on the apparatus in the opposite direction to the wobble and keeps the apparatus from wobbling out of control. The pressure of water then fills in the rest by following a path of least resistance.

The gravitational pull of our Moon on the apparatus causes the phenomenon known as tides. The Moon's gravitational force on the apparatus creates tidal bulges within the Earth's oceans, resulting in the regular rise and fall of sea levels. Looking back at what we've learned about functions, considering these tidal forces, in conjunction with the pressure induced by water- the heaviest combined entity on the Earth, serves as a stabilizing factor.

The gravity created by electromagnetic and centripetal forces increases the pressure of water the deeper and closer the water gets to the source, which is derived from the core. This pressure also functions as a liquid cage. Tidal forces are part of the outside stabilization of the apparatus.

The encased golden double loops are part of an inner cage holding substances (lava) back and slowly allowing them to cool. If this material were to surface all at once without cages to push back, then catastrophes would occur.

The Double Loops and Their Electrical Function

Combined are two single-turn, circular gold loops of electrical conductors, oriented such that they have a common center. As shown in the picture above, loop one lies within a plane and has a counterclockwise current when viewed from the positive north pole axis.

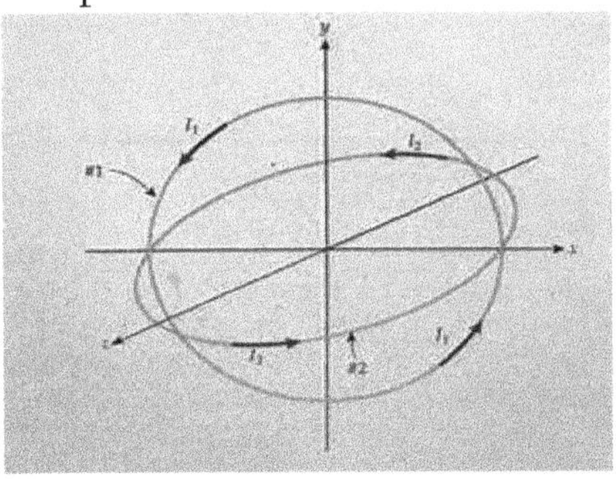

The second loop creates a different plane and has a clockwise current when viewed from the negative flow of the south pole axis. Determined on their equal radius, the magnitude of the net electromagnetic field at the common center creates a constant, continual flow. This magnitude of electromagnetic flow helps produce the electromagnetic field. Let's recap the functions of the Gold Double Loops.

1. They're made of gold.
2. They're combined loops.
3. They support the electromagnetic vortex cage.
4. They act as a conduit and transformer for the continual flow of electromagnetic energy.
5. They provide a track for the gravitational pull of our Moon to follow.
6. They help keep the apparatus from wobbling out of control by guiding our Moon.
7. Along with the gravity of our Moon, they help control tides.
8. They are used to slow the gradient flow of materials back to the surface of our Earth, allowing for cooling.
9. Their electrical function helps produce the electromagnetic field.
10. Overall, they produce a continual flow of energy as a double-loop system.

As the gradients get harder, make sure you didn't miss anything by rereading, studying the pictures, and taking notes.

Just as most structures are built with something to hold them together, the apparatus also has something to hold it together. A cage.

Chapter 13: The Electromagnetic Vortex Cage

The electromagnetic vortex cage is the last element of the apparatus.

Scientists have an agreement that our Earth's vortex is an electrical grid. Just as we study the many functions of our Earth, we also need to study the multiple functions of the vortex grid, which is created by an electromagnetic cage. The cage completes the apparatus.

The cage is a series of metal tetrahedrons. At this point, in order to glide smoothly along a gradient, we must identify what tetrahedrons are and how they function.

The definition of a tetrahedron: Four combined triangles of equal sides.

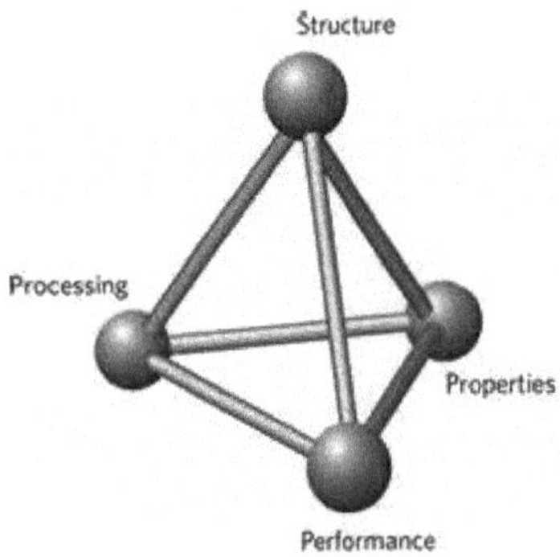

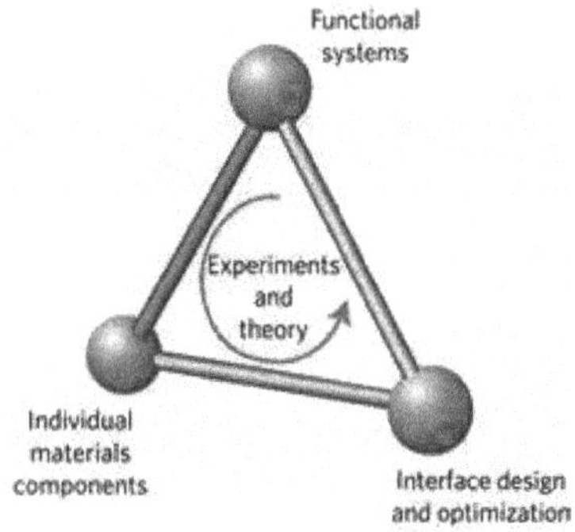

Tetrahedrons, at their points, are electromagnetic balls, and when they are connected, they form a smooth sphere.

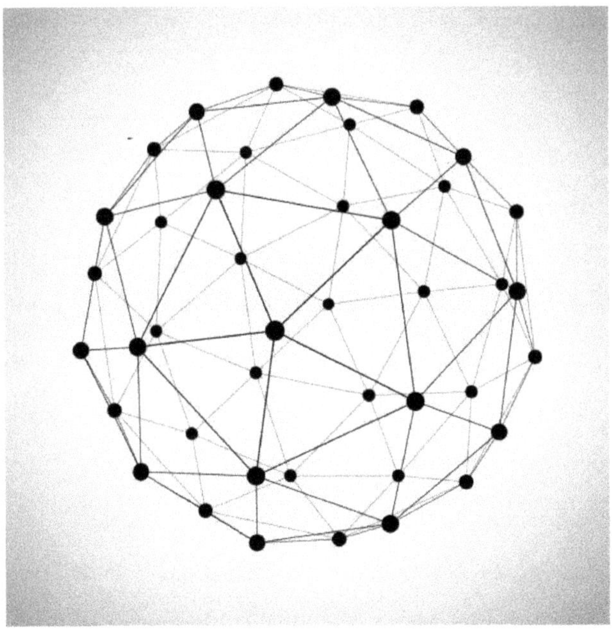

When these points magnetically lock, we can notice they form funnels, one pointing up and one pointing down. This configuration allows for the easy flow of materials.

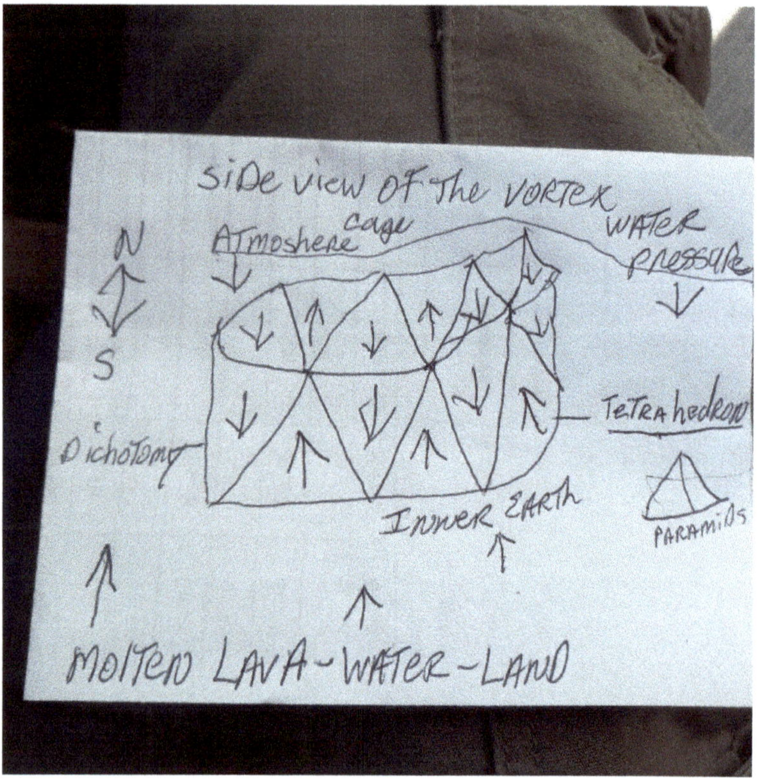

Together, the tetrahedrons, along with the reinforcement of the 'Golden Double Loops,' create an electromagnetic cage. The cage is being electrified by the energy of the positive/negative, north and south poles, or alternators, completing an electrical magnetic circuit.

Together, we can change agreements by just following the gradients of existence. For instance, we can follow the growth of a newborn from birth until adulthood. The agreement is a newborn. The gradient of growth is an adult. When growth is complete, the new agreement is an adult. The gradient of growth is there; we can see it, and above all, we can track the gradient of the individual's life from infancy to maturity.

Look at the pictures above and envision a cage made from solid metals. Then, take note that when you check the oil of your car, the engine is encased for protection and stability. It has side walls and a hood. It is being stabilized with bolts, mounts, and a frame. Of course, Earth's engine is different, yet the notion is the same. Something has to keep it all together.

Look, we can see that the cage is there by observing the pictures of the electrical vortex. Unlike the imaginary line called 'the equator,' there is nothing underneath inside our Earth that shows proof of the equator. The equator does not exist; it is just a false agreement. The equator does not even have a gradient we can follow, as a newborn to an adult or a sapling to a tree.

The cage, or sphere, plays a key role in maintaining the balance and stability of the vortex. The 'Gold Double Loops" serve as the vital connections that hold the cage together. By arranging tetrahedrons in a specific manner and connecting them to their magnetic balls, a fascinating smooth sphere emerges, forming the structural foundation of the electromagnetic vortex cage.

The vortex is simple; it's one of the entities that is holding the Earth together along with the centripetal force of gravity. It's a cage. Centrifugal force would send the elements of Earth, such as rocks, water, us, and fish, flying off into space. The inertia of centrifugal force pushes lava toward the outside of any structure. The cage is made of tetrahedrons or an upside-down funnel.

Due to inertia, the lava is propelled into the tetrahedron, resulting in the emergence of a lava tube. As the lava cools, it shapes land and mountains. Therefore, all solid structures, including Earth's crust, land, and mountains, are formed starting from inside at the core.

Along the center of the system, these tetrahedrons are closer to the surface. The tetrahedrons form islands atolls and islets in Taiwan and Indonesia. The Puget Sound has these same outlets or lava tubes. Something has to hold it together. The other entities are discussed in different chapters of this book.

The vortex is an electromagnetic cage. As we are moving along on our gradients, we are beginning to gain certainty that our Earth didn't just magically form through Mother Nature. The Earth did not form from mysterious rocks that collided billions of years ago, twisting and churning until a certain mother nature thing called gravity began crushing them together. Then small comets with water particles, little by little, formed the oceans, blah, blah, blah! Our Earth is designed and built! Our Earth is a series of mechanical engines combined as one operating apparatus.

Chapter 14: The Mountain Ranges and Volcanoes

Once the gradients take hold while you're reading this book, you will be able to see how mountain ranges are formed. Most everything on the surface of our Earth is created from the apparatus deep inside and is produced by centrifugal and inertia forces. The substances present on the planet, including rock, metals, gases, and more, are grinded and pulverized, then driven down and propelled toward the center of our Earth through the combined influence of electromagnetic and centripetal forces.

Through flighting on the grinder shafts that spin clockwise, the material gets heated, pulverized, and melted. Then, the inertia of the centrifugal force, created by the nuclear core and the electromagnetic discs, sends the heated material back up to the surface. The heated material gets sent along the flighting of the grinders that spin counterclockwise. Then, it cools and forms solids, metals, gemstones, oil, gas, granite, and other varieties of materials. These materials get squeezed through the tetrahedrons of the electromagnetic vortex cage, giving rise to various geographical features such as landmasses, mountains, volcanoes, rolling hills, and valleys.

The trajectory of these grinders contributes to the formation of volcanoes, a path that seismologists refer to as the 'Ring of Fire.' The Ring of Fire is a half-circle of volcanoes along the Pacific rim. They are formed, starting from the Kermadec Tonga Trench outside of Australia, proceeding up into Japan and the Antilles of Alaska, and continuing along the northwestern coast of North America. Then, the 'Ring of Fire' ends along the western coast of South America with the Peru-Chile Trench.

In the earlier chapters, we discussed what engines need in order to function smoothly. One of those entities is the release of combustion or exhaust.

For easy understanding, this area of our world, which seismologists call the 'Ring of Fire,' or volcanoes, is the exhaust system of the apparatus. Basic observations of this area and a simple understanding of engines make this agreement irrefutable. The agreement made earlier that 'Earth is an Engine' shows proof that the 'Ring of Fire' of Earth is an exhaust system in order to release combustion. Moreover, the largest mountain ranges are formed just inside and outside the ring. Proving that a path is being created by **something.** This path is being created by the northern sets of grinders of the apparatus.

If there is a northern exhaust system, there must be a southern exhaust system as well. To prove the balance and equilibrium of a dichotomy, there must be an exhaust system and path created by the southern sets of grinders. Basic observations, again, show proof of a path created by **<u>something</u>** inside our Earth.

Noticing from along the eastern coast of Africa, up through India, and ending along the islands of Indonesia/Malaysia, shows a half circle path of southern volcanoes.

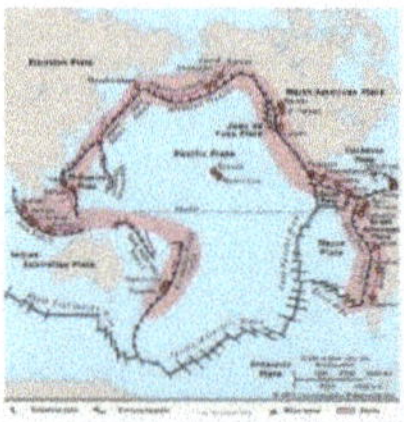

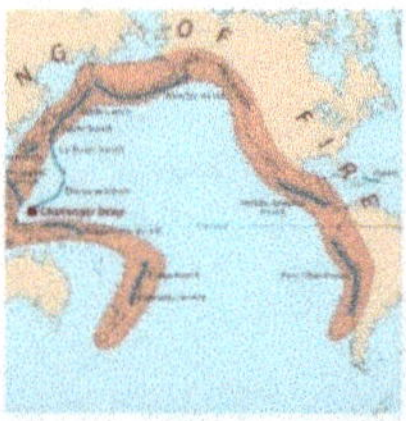

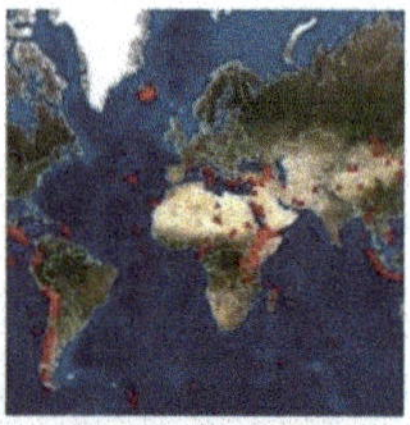

With only three sets of grinders in the southern section of the apparatus, this ring of fire is smaller yet observable. Following the path of the southern sets of grinders along these volcanoes shows they are formed in a clockwise direction, making this observation easy.

These observations become easy when we know what to look for. For instance, one of the easiest mountain ranges to look at is the Rocky Mountains. Start at the northern tip of the eastern side of the Rockies on a map and place your finger around the Aleutian Islands through Vancouver. Then move your finger south and follow the Rockies down through the eastern side of Colorado. Continuing on down, circle through Louisiana and the Mississippi Valley, back up along the eastern side of North America into the Blue Ridge Mountain range. The Blue Ridge Mountain range of Georgia and the Carolinas are some of the most majestic mountains in the world. Continue your line through the Appalachians into Maine, and you will see the path of the northern sets of grinders. You can continue this up-and-down path along the coast of Iceland and the rest of the northern part of the surface of our Earth.

Meteorologists are tracking the four sets of northern grinders inside our Earth as you read this book. They lose track of these giant magnetic balls and conclude they are a myth. The reason they lose track is that they begin to slow down and cool in August through December; this is considered fall. This slowing down and cooling makes them hard to track. When we examine the cause of an early fall, such as September snow in the mountains of Utah, a simple explanation is the northern sets of grinders are shutting down early.

Then, they shut down completely around the winter solstice until March. Both the northern and southern sets of grinders hide their path under oceans and large mountain ranges, making them extremely hard to locate. Reading and following this book will make them easy to find.

Compare the mountain ranges around the world and notice that they are growing, stabilized, or decaying. The Rockies are growing, the Western Sierra's and the San Gabriel's are stabilized. The Santa Monica and Appalachian Mountain ranges are decaying. Centrifugal and inertia forces drive the lighter materials, including water and liquids, towards the surface first, followed by the emergence of magma and other dense substances. The gradual progression of this procedure is evident in the steep yet slow, gradual formation of land. As we explore our Earth with different observations, the evidence of this gradual formation is apparent. A new look will reflect the powerful workings of centrifugal forces in shaping our planet. These forces give rise to majestic mountain ranges. From the oceans to the tops of the highest mountains, we have to agree by now that **SOMETHING** is creating, stabilizing, and causing decline. **Something**, without a doubt, is now an agreement: Earth is a mechanical engine.

Fun fact: Our Moon is the only satellite in our solar system that doesn't have a name or a number. We call it 'Our Moon'.

Chapter 15: Our Moon

The subject of our Moon has been a topic of debate going back to its first observation. By narrowing this never-ending topic down to a function that is rarely discussed, we can provide interesting conversation.

Electrical engineers who design systems will verify that resistors and capacitors in electrical circuits are designed to control the overflow of the electrical input.

Without counteraction or resistance, engines and many other entities would operate out of control, even us humans. Without pain receptors throughout our bodies, we would be completely damaged as beings.

Therefore, the main function of our Moon is to act as a capacitor or resistor, which in turn prevents our Earth from wobbling and spinning out of control. Amongst other effects, our Moon provides equilibrium and stability.

Scientists claim that over time, our Moon formed from loose rock and became solid by gravity. What exactly the Moon is made of can be determined on a different plane, at a different time, that surpasses the gradients of this book.

One thing I'm going to stress is that our Moon absolutely cannot be drilled into or mined for its steel or for any reason. The devastation of this celestial sphere will result in the destruction of Earth. For instance, removing a piston from an automobile engine and giving the design a different function makes the engine unstable. Total destruction of the engine will occur, and a new design will undoubtedly be required.

The basic thought on what our Moon consists of is that it has an electromagnetic charge similar to Earth's. Since I'm not a scientist or physicist, nor do I claim to be any of the sort, this is only theory. Therefore, what our Moon is made of is not as important as how it functions. Our Moon functions as a counterweight. It stabilizes our Earth and keeps it from wobbling out of control and spinning off into space. Overall, our Moon is a dynamic force that pushes back and continues to provide equilibrium and balance.

First, let's take a look at our Moon's orbit. The orbit of our Moon adequately demonstrates that, in itself, slowly spins counterclockwise, just as our Earth, overall, spins counterclockwise. Therefore, there must be further examination of the interrelationship between our Earth and our Moon, which serve as counterweights operating in tandem. Just as cogs jointly participate in a mechanical machine.

The orbit of our Moon follows two circular magnetic tracks inside Earth consistent with magnetic metals. These tracks were discussed earlier. They are the encased Gold Double Loops.

The Moon is constructed of metals that are proficiently magnetized toward the electromagnetic vortex cage. Inside the electromagnetic vortex cage are gold double loops. The electromagnetic vortex cage maintains our Moon in orbit, while the Gold Double Loops guide the orbit of our Moon.

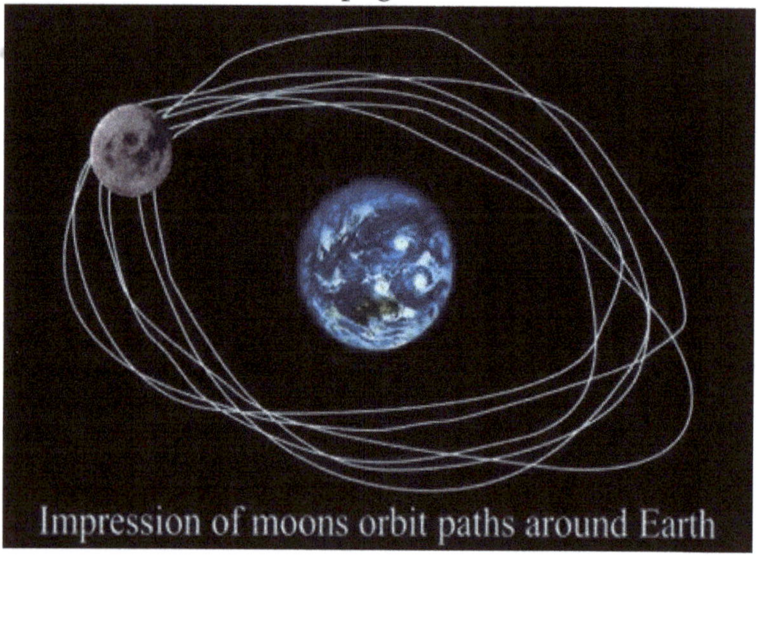

Impression of moons orbit paths around Earth

When the Earth slowly wobbles out of position, our Moon begins acting as a counterweight, creating and maintaining equilibrium. Our Moon acts as a guiding force, steering Earth back into balance. The Gold Double Loops act as a circular track, providing the systems of our Moon with stability. Studies show adequate proof of this by viewing the previous pictures of our Moon's orbit.

Within the logs of the early years of navigating on the oceans and seas, captains studied the celestial skies in order to find better routes for trade. The fool of a captain would be the one that followed the Moon. The early captains of the seas studied the skies above instead of the phenomenon occurring below in the depths of the Earth. The same to this day, we study effects from above instead of the cause from below. Relying on our Moon for navigation could potentially guide a ship in an endlessly circular trajectory across the oceans without ever reaching land. This is why we can see our Moon and our Sun in the same sky. Look at pictures of other moons, and you will find that they are smooth spheres. Then, look at Phobos and Deimos, the moons of Mars.

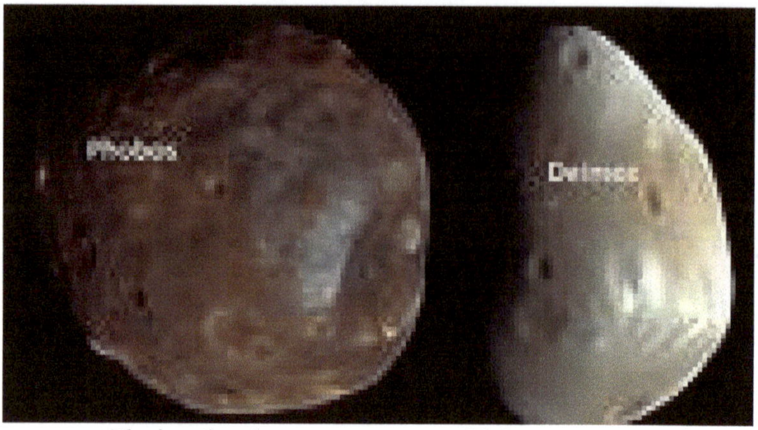

Obviously, the two moons of Mars are not spheres. I'm pretty sure that whatever happened to Mars' one Moon didn't turn out well for the planet. I'm also sure the same will happen to us if we allow our Moon to be drilled and sold for profit.

Huge deposits of metal have been discovered at the bottom portion of our Moon. One thing the people of Earth must never allow is the excavation and drilling of our Moon. Our Moon stabilizes the apparatus inside our Earth, and any derivations to the sphere of our Moon will, without a doubt, destabilize our Earth.

Chapter 16: Our Sun

Certain things are easy to identify about our Sun. Our Sun provides light, heat, and essential elements for life. The main topic of this book is about Earth; what needs to be conveyed about our Sun has to be described in brief. The influence our Sun has on our Earth is only important to describe its similar dichotomies. For instance, scientists agree that our Sun doesn't have moons. Our Sun absolutely has moons. The functions of Earth's Moon verify with certainty that our Sun has moons.

This is easily proven by simple observation. Studying the functions of other moons in our solar system proves that the moons of our Sun are placed directly where they're supposed to be. They systematically act in a similar fashion as most other moons. They act as stabilizers, holding our Sun in place and keeping it from spinning and wobbling out of control. Mercury and Venus are the moons or stabilizers of our Sun; they operate together. While Venus spins clockwise and Mercury spins counterclockwise, they form a balanced dichotomy. Studying the orbits of these two spheres around our Sun makes this observation extremely obvious. The operation and functions of our Sun, what it's made of, its size, and how powerful its gravity is, create a different topic set for another book. Our Sun is a completely different engine than our Earth. Our Sun has no need to keep cool, yet because it is spinning, it still needs to be stabilized. This engine is by far the most powerful engine in our solar system.

Just like all engines in our solar system, our Sun needs stabilizers, balance, and equilibrium. Stabilizers are what we call moons. In order to keep our Sun in position, it has to be stabilized. Unlike our Moon, we have names for our Sun's moons. The stabilizers are called Venus and Mercury.

Mercury and Venus are the only planets in our solar system that do not have moons. Surprisingly, their lack of moons serves as a unique characteristic, as they themselves play the crucial role of stabilizers. This intriguing dichotomy is further reflected in Mercury and Venus's distinct opposite direction of spin and their orbits- Mercury spins counterclockwise, while Venus spins clockwise.

This seemingly opposing rotation pattern creates a delicate balance and stability for the entire solar system, especially in relation to our Sun. Their orbits and interactions with our Sun contribute to our entire solar system moving through the Milky Way Galaxy. They act as counterweights. Venus and Mercury have been studied as planets. As our solar system travels through our galaxy, research shows that the solar system is being held together by the gravity of our Sun. We must ask the question, "What is keeping the planets and our Sun in place and from wobbling off course? '*Mother Nature*'?

Doubtful!

Gravity?

Partially.

Counterweights and stabilizers we call 'moons.' Plus, nuclear linear gravity. Without a doubt, the engine of our Sun remains a mystery.

It is hard to tell that our Sun created itself by coincidence and over time. For right now, let's put what the engine of our Sun is made of aside and look at what it does. Along with keeping us warm and helping plants grow, plus a thousand other facts, one very important observation we all must have missed, which can be narrowed down to one word, is 'Slingshot.'

Our solar system slingshots around our Sun.

Earth has the number one slot in our solar system because Mercury and Venus are stabilizers. Next are Mars, Jupiter, Saturn, Neptune, and Uranus. We can call the planets whatever we want. Knowing how they function is absolutely more important. Every time we slingshot around our Sun, we travel through the galaxy. Astronomers propose a theory suggesting that we are being drawn toward a colossal giant black hole located at the center of our galaxy. My guess, however, leans towards the insight of another gravity engine.

Let's look at what we have in front of us. We have engines called planets that spin and rotate around our Sun.

When the grinders begin to warm up, plants begin to grow. Bugs begin to crawl and hatch, bears begin to stir, and birds head back north. How can this be only related to our Sun when March is colder than November?

In November, the Northern engine is getting colder, and in March, it's warming up. Listen to what people talk about in November when they talk about the weather. "It's getting cold; it's going to be a tough winter." Then, in March, what do you hear people say, "It's warming up!" A fun fact statement would be,

“March: in like a lion and out like a lamb.” Most of the time, anyway.

What’s getting cold, and what’s warming up?

Our Sun?

No.

Mother Nature?

Absolutely not.

Yes, the Earth is tilting back and forth on its axis, causing our Sun to be closer and farther away, depending on where you are on Earth. There are, of course, other functions. The creation of a clear blue sky in the middle of June is rarely studied by scientists. It doesn't make sense.

Weathermen, astrologers, and meteorologists stop studying causes and just report effects.

Fun fact: Why do leaves fall off trees?

Fact: Our Earth tilts back and forth on its axis.

Fact: Our Earth is an engine, and just like most engines, the apparatus inside Earth needs to be kept cool in the summer and warm in the winter. Therefore, leaves begin to grow in spring, providing shade to keep the surface of our Earth cool. The leaves begin to fall off in the fall, providing more sunshine to reach the surface of Earth, keeping our Earth warm in the winter.

Let's have some more fun and talk about the weather.

Chapter 17: The Weather

The first thing we need to start with is that all weather conditions are effects. Knowing this, let's look at how we measure the readings for the temperature of the weather. A long time ago, and at about the same time climate change alarmists began their argument, the measurements for the temperature of weather changed from using a poisonous metal called Mercury to digital. Now, the temperatures of the weather are measured with precision by using digital technology. This fact is never used when engaging in discussions about climate change. Simple question, "When should we begin to compare figures from the past, and which measurements should be used for comparison?" Measurements of temperature from the past should only be used from the beginning of the switch from Mercury to digital. For instance, the temperature from a certain area measured in the location of XYZ in 1988 using Mercury should be filed for comparison in a different category.

The measurement of temperatures is going to be more precise using digital. Unless the past measurements of temperatures before using digital are archived, the exact figures will be debatable.

The formation of weather is not a cause. As weather forms, it is an effect. After weather happens, such as the destruction of a tornado, the destruction is the effect, and the cause of the destruction is the tornado. Simple question, "What caused the weather and tornado?"

When we look upon a cloudless, sunny sky, with the Sun shining brilliantly, does weather occur? What is the cause of a bright, splendid sunny day?

Something

I've heard a weatherman say, "It's not what I saw; it's what I didn't see." When we only study what we see, we leave out the most important results. The results of the effects are only half of the dichotomy. This is part of the problem, the problem of intellectual setbacks created by only studying threatening or dangerous weather. Only studying effects, again, prevented the scientific community from figuring out the cause. We should all realize that the weather is not exact. A quick note to all the people who study the skies, the stars, planets, and the beautiful entities that strike our imagination: they aren't exact. When there are so many things that aren't jointly connected and are spinning in different directions, finding common denominators will be extremely difficult. One thing is for sure: Obviously, they didn't just form on their own over time.

When some meteorologists (weathermen) finish their schooling, they also finish their further studies of the weather. Then, they not only become television reporters but also contribute written pieces on weather-related outcomes. These reports focus only on the effects.

Take fog, for instance. Hours or days before a fog settles in or moves across a section of our Earth, it is muggy, hot, and steamy. Then, the fog brings cool air. This is all effect; meteorologists study the fog and not the cause. You'll find results of these studies, reporting that hot air rises, meets cool air, etc., which are studies of effects, again.

The thing is that hot air, called steam or humidity, flows out of the ground because an electromagnetic set of grinders, rotating counterclockwise, just passes beneath the surface of the Earth.

These sets of grinders are the cause of most weather.

The grinder caused intense heat, and the heat reached the water level of the Earth. Steam was produced inside of Earth, and by the time it reached the surface, the steam began to cool, creating the effect called fog. Take a frying pan, place it on the burner of a stove, and turn the heat up full. When the pan gets hot, pour a glass of water on it and watch what happens. What you will witness is an effect; the cause is not the pan. The cause is the energy being produced from the stove.

Continuing on our observation, we found out the actual cause was electricity or gas. Therefore, what produced the electricity and gas? Following the same process, we should ask,

"What causes the weather?" When researching the answer to this question, all we find are more effects. Answers, such as hot air meets cool air and blah, blah, blah, effect, effect, effect.

Again, what causes hot, beautiful Sunny days?

Something.

Yes, **something** causes a beautiful day.

Mother Nature again?

Doubtful.

What about a cloudy day or a rainy day?

What about snow storms, tornadoes, hurricanes, lightning, floods and wind? Mother Nature is the agreement. Does this agreement make sense? How about we agree that **something** caused the weather, therefore the weather is an effect? When we look into the sky at the weather, we see the results of **something** at cause, creating effects.

No matter what we do, no matter what we are surrounded by, we have to use a common denominator investigation on the effect called weather.

Finding a common denominator is not that simple since the weather is being created by many different structures of the apparatus.

Something causes the weather. Scientists use the phrase, "Wherever there is an action, the result is a reaction." Again, when you look at the weather in the sky, what do you see? An action or a reaction, or do you see the cause, or do you see the effect?

When meteorologists and scientists go up to the clouds and study them, they find that there is nothing up there. If you go to the clouds, you'll find nothing but clouds because it is all effect. All you have to do while you're up there is look down, and you will see **something**.

This **something** is called Earth. Then, when you plant your feet on Earth again, you won't see anything that is capable of creating weather. Then, look down again, and simple observation tells you that the cause is being created by

something inside of Earth.

Overall, the engines inside of Earth are the cause. It's not the heat or the atmosphere that is zapping all the water out of the sky; the atmosphere, once again, is an effect. The atmosphere doesn't require water; it's the engines that require water. When we get to the cause, then we can understand the effect. When we know what is causing the effect, we can continue to analyze it and develop a deeper understanding. So, what is causing the effect?

"*Mother* Nature?" Doubtful!

How about Earth is an Engine?

I hope your mind has expanded and we started a new agreement and gradient. A gradient that far exceeds one you have ever experienced.

I can't just make the statement, "Earth is an Engine," without proof.

Well, look around. The Earth spins counterclockwise. The Earth creates an electromagnetic field. We have ocean currents and blue holes along the Bahamas and Bermuda.

We have volcanoes, hurricanes, tornadoes, and lightning; our Earth tilts back and forth on its axis. We experience the winter solstice and the summer solstice. The four seasons- winter- spring- summer, and fall- occur around the same time every year.

Fog and humidity clearly are reactions to something at cause.

In San Francisco, they even give names to these effects: they call the fog 'Karl' and little eddies.

We can watch or listen to the weather report, and there is no denying that fog is being created by cold air, meeting hot air and barometric pressure. These are all effects and reactions. We have to agree by now that the cause is designed. Therefore, the cause of heat pouring out from the surface of our Earth is being produced by the engines of our Earth. We can follow the path of the sets of grinders that create the weather patterns of North America. Let's follow the path of just one set of northern grinders. Notice the creation of land and mountain structures along the way, starting from the Aleutian Islands. Swinging south under the Rocky Mountains, barreling down through northern Texas, then swooping back up around the plains of the Mississippi Valley and continuing northeast of Tennessee, Kentucky, and the Carolinas, these grinders create the Smokey, Blue Ridge and Appalachian Mountain ranges. Then, they pass under the New England states and head out under the Atlantic Ocean. We can identify this path by the creation of the state of Maine. We can continue to follow the same swooping weather patterns of the North throughout Eurasia.

Just watch the weather channel. Watch the spinning spheres on the screen, one circle with an "H" followed by one circle with an "L," swooping down and then up.

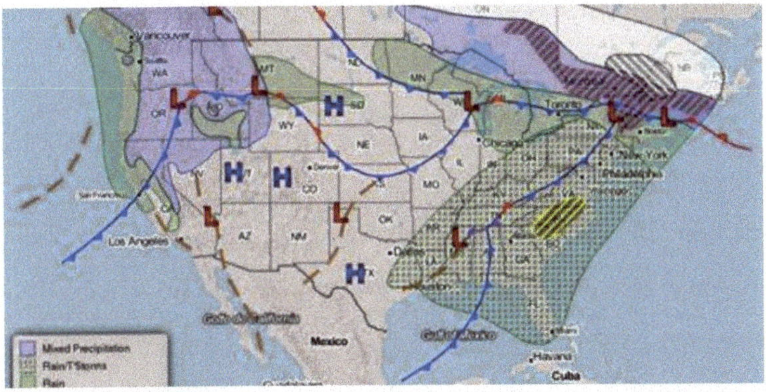

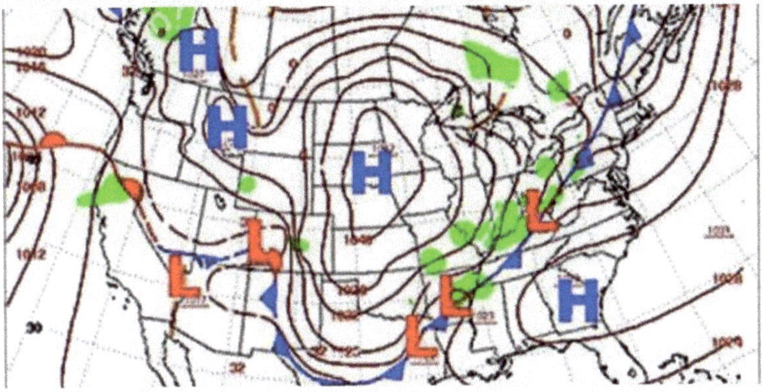

Lightning storms in the northern part of the Earth start in the Middle of March and end in the middle of September. This statement is proven by noticing that the Cumulonimbus clouds eventually go away completely during this period of time. We can also observe different weather patterns and cloud formations from the fall months until the end of February and the beginning of March. The reason is that the grinders are downshifting and slowly shutting down until spring. Remember the dichotomy that the opposite is happening in the South.

Lightning is an effect, then becomes a cause, creating thunder. Lightning is the effect from the cause of the apparatus and grinders producing electricity. The grinders are the main cause of lightning. They project electricity and microwaves through the tetrahedrons of the electromagnetic vortex cage. This combined heat and electricity produces an effect. This effect is called Cumulonimbus clouds. These clouds are only produced while the grinders are spinning. The Cumulonimbus clouds fill with heat, electricity, and then water. The main effect is thunderstorms. The set of grinders, as it passes through the Earth, scorch the land. The extreme hard rain produced by the thunderstorms cools our Earth. Just like pouring water in a hot frying pan, steam is produced. This water is collected back into cumulus clouds, which produce a softer rain.

Studying effects and trying to change agreements is something that needs to happen in order to create a better understanding of how the Earth works. Follow the spheres of the grinders as they begin their journey from their resting and cooling place in the North.

Let's unravel some fascinating mechanisms behind the Earth's electromagnetic field having a significant influence over the effects called weather.

The sets of grinders emit microwaves and electricity through the electromagnetic vortex cage, a phenomenon evident in pictures of cumulonimbus clouds. These clouds exhibit massive mushroom formations. These formations of clouds are formed by the energy shooting through the tetrahedrons of the electromagnetic vortex cage, forming different clouds in the sky. The Cumulonimbus clouds are linked to massive lightning storms. In the North, the grinder activity gradually slows down around the middle of September to the beginning of October, leading to a complete disappearance of cumulonimbus clouds until the end of February and the beginning of March. As the northern grinders begin to warm up around February/March, winds pick up, and April showers follow, bringing May flowers.

In the South, the southern sets of grinders slow down. At the same time, they simultaneously coincide with the acceleration of the northern sets of grinders. This opposition produces significant forces during the fall and spring seasons. These climatic shifts give rise to Hurricanes in late summer and fall. Tornadoes produce their destructive force in late winter, spring, and early summer. This gradual gradient of yearly cycles reveals an even more intricate working of our planet's weather. They unveil a devastating interplay of forces that shape powerful dichotomies across the globe.

Again, the forces of tornadoes and hurricanes are being produced by the dichotomy of one set of grinders turning on, speeding up, and the other set of grinders slowing down and shutting off around the same time every year. The turning on and shutting off of this opposite spinning also creates the tilting of our Earth's axis.

At this point of our long gradient, we can at least ponder the fact that the Earth was constructed by God and his crew of Angels instead of just being the result of Mother Nature.

Next, let's talk more about hurricanes and tornadoes.

Chapter 18: Hurricanes and Tornadoes

Compare pictures of hurricanes with some pictures of our Milky Way Galaxy.

Within the Earth, when the four sets of grinders in the North are warming up, beginning to spin and grind, the centripetal and centrifugal forces of the apparatus begin to strengthen. Along with the forces of the three sets of grinders in the South, which are shutting down and cooling off together, they collectively generate an extreme counter-reaction force along the real equator. These combined forces are called tornadoes in the North, the cause being all seven sets of grinders spinning in opposite directions at once. The effect of the grinders and centripetal force are acting like a counter-reaction. This counter-reaction is called tornadoes and cyclones in the North.

The continuous counterclockwise spin of the main nuclear engine, along with the electromagnetic discs, is constant along the real equator.

Of course, this action/counteraction/reaction forms a balanced dichotomy. The result of these forces forms a balanced opposition. Unfortunately, even though these forces create balance and equilibrium, societies suffer from the destruction they cause. As we follow the track, or the effect called a tornado on the surface of our Earth, and we search for a cause, finding a simple observation beneath the surface of our Earth becomes easy. Deep inside the Earth, we can follow the cause of the sets of grinders.

There is a specific geographic region known as Tornado Alley, located in the central plains of the United States. Within this region, the interaction between the counterclockwise rotation of the northern grinders and the forces of the real equator battle the opposite spin of the southern grinders. Tornado Alley is the battleground of these combined different forces, creating devastation year after year. This interaction involves a downward atmospheric pressure and is one of the key factors of this effect. Specifically, just one of the four northern sets of grinders, spinning clockwise with their flighting spinning down, creates a low pressure. At the same time, one of the southern sets of grinders in close proximity spinning counterclockwise, with its flighting spinning up, causes a high-pressure system. Then, combine the forces of the main nuclear engine and the two magnetic discs along the real equator, and we have catastrophes. When these opposing forces clash, tornadoes are formed.

The battle between the forces of the northern sets of grinders speeding up, along with the southern sets of grinders slowing down, and the forces of the real equator generate the conditions for the formation of tornadoes during the early months of the year. The opposite occurs at the end of summer or early fall, creating hurricanes as the northern sets of grinders slow down, and the southern sets of grinders begin to speed up. Understanding these three combined extreme interactions sheds light on the dynamics of the engines below, offering valuable insights into what causes tornadoes and hurricanes.

When we track and attempt to study tornadoes and hurricanes, since all of Earth is operating on a dichotomy basis, we are only studying half of the formula called 'Cause creates an effect.' We are studying the effect without studying the cause.

The movement of water holds a crucial role in shaping the intensity and frequency of weather events. We need to consider the intense forces being created by the engines of the apparatus. The impact of these powerful engines plays a significant part in drawing water down through centripetal and electromagnetic force. These extreme forces are used in order to cool the apparatus, along with other various sources. The only main cause of weather, amongst many other elements of weather patterns on a grand scale, is to keep the engines of the apparatus cool. As mentioned before in the chapter on engines, in order for an engine to tilt a planet back and forth on its axis, takes immense force. The cause of this force creates tornadoes, hurricanes, and many other disastrous weather conditions. These conditions will only get worse unless, together as a society, we learn how to keep the engines of our Earth cool and warm and balanced with equilibrium.

Although these weather effects are extremely devastating, based on the violent causes taking place within our Earth, these balancing effects are quite manageable. One fascinating example of this interaction is seen in blue holes found in places like Bermuda, the Bahamas, and other locations around the world. These blue holes are indicative of the location of this central engine, where water is drawn in, acting as water intake valves and producing oceanic phenomena.

Understanding how this main engine influences the availability and movement of water is essential for comprehending the complexities of weather events.

Now, it's time to finish this gradient of a book and move on to the next one. Before I go, I would like to leave you with solutions to,

'Climate Change.'

Chapter 19: Climate Change

Here are the answers to climate change: nine words structured into two agreements or sentences.

<u>1. Filter</u> it.

<u>2. Oil</u>, or **<u>Goop</u>** out, **<u>Goop</u>** back in.

How about that for changing the world?

Let me explain by asking a simple question. "How can scientists all around the world be so close, yet so far away, from the solutions to climate change?

Carbons in the air serve the purpose of keeping the trees growing and keeping everything green. Some CO_2 is produced by anything that breathes in the air, and during exhaling, the lungs convert the air from oxygen and a mixture of nitrogen and hydrogen into CO_2. If too much CO_2 and carbon in our Earth's atmosphere is the real problem of climate change, and scientists all over the globe say that it is, then the answer is simple.

Filter the CO_2 and carbon out of the air.

Can scientists do this?

Sure, during the famous return of Apollo 13, after an explosion blew off the side of the main ship, the astronauts had to move into the moon lander, shutting down power to the main ship. The moon lander craft was only built for two astronauts. Since there were three astronauts, CO_2 began to accumulate on the trip home after sling-shooting around the Moon. The astronauts built a filter out of cardboard, hoses, and tape. This procedure saved their lives, amongst other heroine feats. As we examine the major cities around the world, we can notice that most of the pollution arises from the general masses of people in one area.

Beijing, Hong Kong, Moscow, New York, and Los Angeles produce the most carbon in the air. Some of these cities are surrounded by mountains. The area on the mountainside that is inhabitable can easily be transformed into giant air filtration systems. New York has skyscrapers taller than most mountain ranges. What would be so difficult than to build huge filtration systems in the mountains surrounding these great cities, right where the smog layer lingers? Let's convert the thirteenth floor of the skyscrapers into carbon filtration units.

When we go inside these tall buildings, there are air filtration systems that intake the air from the outside, filter it, and then pump the air into the building so people can breathe. Yet, on the other side, they don't filter the air going out. An easy design for these buildings would be to filter the carbon out of the air that is exiting the ventilation systems.

Let's design a house with the top floor having a small induction wind turbine. As wind flows through, it filters the carbon out of the atmosphere and, at the same time, creates electricity to operate the home.

This, again, is not the perfect solution.

This would lower the energy used from the power grid and the use of oil. That's simple. I'm going to explain to you why this is only a small solution to the problem of climate change.

The title of this book is 'Earth is an Engine.' We went over the simple needs of Engines.

The main reason engines function is because they are designed to stay cool.

I'm not saying that scientists are wrong about trees producing oxygen, wood for burning and building, and foundations for the homes of animals. The habitats within jungles and forests are responsible for generating the majority of our Earth's oxygen and shade. Therefore, as stewards of future generations, we must guard these ecosystems from deforestation. The agreement is that we have to include a main feature of shrubs and ground cover, and that is shade. Because these entities keep our Earth cool in summer and warm in the winter. Whatever happened to us using less paper by 'Saving a tree saves the planet?'

Oil is the main substance that keeps the engines lubricated, reducing friction and keeping the engines cool. Water is the most abundant coolant on the planet. Imagine a world where we harness the filtration systems on a grand scale, revolutionizing how we approach one of our planet's most pressing challenges.

The cost of filtration units is relatively reasonable, amounting to a few billion. However, the true answer lies in the role of oil, or 'GOOP out, GOOP back in.' A simple experiment of draining oil from a car and replacing it with water reveals the crucial function of oil as a lubricant. Earth itself produces its oil, but over the past few hundred years, we've been extracting and depleting it and storing it for reserves. When we remove oil from our Earth, it leaves an empty hole or cavern. Then, oil companies replace the oil or fill the empty cavern with water and sand. Fracking is allowing oil companies to go deeper into our Earth than ever before.

Instead of replacing the removed oil and filling the caverns with sand and water, we, as humans, without a doubt, need to replace the oil with a substance that produces the same result as regular oil.

The need arises to consider alternatives like synthetic oil lubricants, which mimic the properties of oil.

Though deforestation and other issues are concerning, the removal of oil remains a significant and pressing challenge to address for the well-being of our planet. The problem of climate change is not only the burning of oil; it's mostly the removal of the oil from the engines of Earth.

OIL, OR GOOP OUT, GOOP BACK IN.

This concludes our extreme adventure.

Agree that Earth is an Engine, so we can,

'Save our engines, save our Earth, save Humanity.'

www.ingramcontent.com/pod-product-compliance
Lightning Source LLC
LaVergne TN
LVHW051045210926
848486LV00001B/26